T0207380

# Problem Books in Mathematics

**Series Editor**

Peter Winkler
Department of Mathematics
Dartmouth College
Hanover, NH
USA

Books in this series are devoted exclusively to problems - challenging, difficult, but accessible problems. They are intended to help at all levels - in college, in graduate school, and in the profession. Arthur Engels "Problem-Solving Strategies" is good for elementary students and Richard Guys "Unsolved Problems in Number Theory" is the classical advanced prototype. The series also features a number of successful titles that prepare students for problem-solving competitions.

More information about this series at https://link.springer.com/bookseries/714

Paolo Toni • Pier Domenico Lamberti
Giacomo Drago

# 100+1 Problems in Advanced Calculus

## A Creative Journey through the Fjords of Mathematical Analysis for Beginners

Springer

Paolo Toni
Liceo Scientifico Statale Enrico Fermi
Padova, Italy

Giacomo Drago
Istituto Tecnico Economico Calvi
Padova, Italy

Pier Domenico Lamberti
Dipartimento di Tecnica e Gestione dei
Sistemi Industriali (DTG)
University of Padova
Vicenza, Italy

ISSN 0941-3502          ISSN 2197-8506   (electronic)
Problem Books in Mathematics
ISBN 978-3-030-91865-1          ISBN 978-3-030-91863-7   (eBook)
https://doi.org/10.1007/978-3-030-91863-7

Mathematics Subject Classification: 26A06, 26A15, 26A24, 26A27, 26A42

This Springer imprint is published by the registered company Springer Nature Switzerland AG
The registered company address is: Gewerbestrasse 11, 6330 Cham, Switzerland

# Preface

This is the revised, translated edition of *Esplorando l'Analisi Matematica*, by P. Toni and P. D. Lamberti, which was published in Italy in 1996. The new evocative subtitle, "A creative journey through the fjords of Mathematical Analysis for beginners," pays homage to the original 1994 manuscript and reflects the original vision of the authors. The premise is to take the reader on a relaxing journey through sets and functions, thereby illustrating the basic principles of Mathematical Analysis.

This revised edition, while remaining quite faithful to the original book, also includes new material concerning the theoretical background necessary to understand and solve the problems presented. A few chapters have also been merged and reorganized.

The book is divided into 8 chapters covering classical topics in Mathematical Analysis. Chapter 1 is a brief review of the basic properties of inequalities which are used throughout the book. Each subsequent chapter begins with an introductory section in which relevant definitions and theorems are provided for easy reference. Problems are then presented, the solutions of which are described in detail at the end of each chapter.

Chapter 2 is devoted to introductory notions of point-set topology on the real line, with special attention to accumulation points, to supremum and infimum of sets and sequences, as well as to the notion of function and its basic properties.

Chapter 3 is devoted to the notions of limit and continuity of a real-valued function defined on an arbitrary subset of the real line.

Chapter 4 focuses on differentiation.

Chapter 5 is devoted to classical theorems of differential calculus, such as those of Rolle, Lagrange, Cauchy, and de l'Hospital (in particular, it contains a geometric interpretation of the corresponding proofs).

Chapter 6 is devoted to classical applications of differential calculus with respect to the study of monotonicity, maxima, minima, convexity, and inflection points of functions.

Chapter 7 pertains to the study of graphs of functions; this chapter is divided into two parts. The first part is devoted to the problem of plotting the graph of a given function. The second part is devoted to the "inverse problem," which consists of finding a formula that approximately represents a given graph (e.g., such that it has the same asymptotes).

Chapter 8 is devoted to Riemann integrals and antiderivatives.

Each chapter is almost self-contained and would be appropriate for students attending a standard course in Advanced Calculus. Furthermore, we believe that this book could be used, on its own, as the basis of an entire course—or as a "course backbone" which instructors could utilize along with additional material, as required for a specific curriculum. (For instance, with respect to point-set topology, we do not discuss the notions of the boundary and interior of a set, or other notions that would better be discussed within the general framework of metric spaces.) This book, as its title indicates, could certainly be used for a standard lecture course in Advanced Calculus. More specifically, depending on the specific course syllabus, one could say that it pertains to Advanced Calculus I (keeping in mind that Advanced Calculus II is often devoted to calculus in several real variables, a topic not discussed in this book). Depending on the level of difficulty desired, in principle, it could also be used to provide supplementary material for an introductory lecture course in Calculus. In some European countries (such as Italy), this book would be excellent for a first-year lecture course in Mathematical Analysis for the bachelor's degree in mathematics, physics, or engineering.

Please note that a discussion of some pedagogical ideas implemented within the book can be found in the paper by G. Drago, P.D. Lamberti, and P. Toni, "A bouquet of discontinuous functions for beginners in Mathematical Analysis," Amer. Math. Monthly 118 (2011), no. 9, 799–811.

The first and the second author are deeply indebted to the third author who produced all of the pictures for the revised version and provided the solution to the new Problem 41*. Without his immense patience and excellent computer skills, this project would not have been completed.

Padova, Italy                                                                    Paolo Toni
Vicenza, Italy                                                      Pier Domenico Lamberti
Padova, Italy                                                                  Giacomo Drago

# Contents

# Introduction to the Journey

We are about to embark on a journey. Imagine, if you will, that we are on a silent, smooth-running boat, and that we are preparing to enter some of the most beautiful "fjords" in Mathematical Analysis. A fjord, you will recall, is a long, narrow, and deep inlet of the sea, between high cliffs. In this book, since we are exploring mathematical "terrain," the cliffs are immense mathematical structures, stretching higher than we can imagine. As we cruise between and beneath them in our boat, deeply entering these mathematical "fjords," we will be able to see the underlying foundations of the massive mathematical theories and structures above us.

The key foundations, or "ports of call," of Mathematical Analysis are not as numerous as the unsuspecting tourist might think. On our journey, we will have the opportunity to pass near to, and thus to closely examine, the fundamental concepts of *function*, of *infinity*, of *limit*, and of *differentiability*. In order to "visit" these "destinations" in a complete way, however, we will have to look at them from many different vantage points; we will have to explore them in some really unexpected ways. It is necessary to allow them the possibility of expressing their full potentials, without making excessively restrictive assumptions.

Keep the image of an unexplored coastal shoreline in mind. If we look at the shore from our vessel, while we are still some significant distance away from the land, it often resembles a straight line. But as we approach more and more closely, we may discover that it has many irregularities and inlets. Similarly, in order to appreciate the fine details of the whole mathematical picture, it is important that we set our imaginations loose as we simultaneously explore with more and more precision.

To successfully achieve such exploration, mathematically speaking, it will be necessary to remain faithful to the definition of *function* as a relation between sets. In fact, a function is not always a mere combination of symbols and operations that transform one set into another. Formulas or equations do not exhaust all possibilities; functions can be much more general and may exhibit fantastic and incredible properties.

By committing to this approach, the notion of set returns to the forefront of our exploration; the study of sets and their topology becomes crucial. We will see that the "Pythagorean-Euclidean drama" concerning the notion of *point* naturally makes its appearance, with startling consequences related to the idea of infinity. For, if we agree to drop the basic classical notions discussed in textbooks, it turns out that the idea of mathematical infinity furnishes an unfathomable wealth, and can play remarkable—indeed, infinitely many—tricks on our intuition.

Another aspect of our journey will revolve around the limits of functions defined on arbitrary sets and not merely intervals. This will provide us with a "window" through which to view inaccessible "terrain," and scenarios playing out on that terrain, which it would otherwise be impossible for us to encounter on our voyage. This part of our journey will give the notion of limit the central role it deserves in the program of Mathematical Analysis.

The notion of *continuity* also yields many surprises and is another fertile plain that we will have the chance to survey on our journey. Indeed, while the definition of continuity appears very simple when we associate it with the idea of an uninterrupted wire, it is very suggestive and can even prove quite tricky when it is more fully revealed by the Squeeze Theorem. (Not to mention the problem of *tangency*... a "caress" produced by the derivative.)

In order to give readers the greatest chance of navigating through the mathematical landscape in a relaxed, enjoyable manner, the book includes some additional material on the order properties of real numbers, the topology of the real line, the definition of limit, and a few graphic features of algebraic curves, as well as the notion of integral.

The book proposes a series of problems which require a bit of imagination and critical thinking. Readers will have the opportunity to compare their answers with our solutions, and to then proceed in journeying down these fjords on their own—possibly forging new paths and discovering uncharted inlets worthy of admiration and further exploration.

# Notation

| | |
|---|---|
| $>$ | Larger |
| $\geq$ | Larger or equal |
| $\longrightarrow$ | Implies |
| $\longleftrightarrow$ | Is equivalent |
| $\lightning$ | Contradiction |
| $\mathbb{N}$ | Natural numbers including 0 |
| $\mathbb{N}_0$ | Natural numbers excluding 0 |
| $\mathbb{Q}$ | Rational numbers |
| $\mathbb{R}$ | Real numbers |
| $[a, b]$ | Closed interval (that is $a \leq x \leq b$) |
| $(a, b)$ | Open interval (that is $a < x < b$) |
| $f : [a, b] \to \mathbb{R}$ | Real valued function defined on $[a, b]$ |
| $D(f)$ | Domain of the function $f$ |
| random$(x)$ | Random value depending on $x$ |
| $\log x$ | Natural logarithm of $x$ |

# Chapter 1
# Summary of basic theory of inequalities

## 1.1 Theorems on the order properties of $\mathbb{R}$

In this brief chapter, we recall a few theorems on the order properties of real numbers that will be used in the sequel.

## 1.2 Theorems on the sum

$T_1$ For all $a, b, c \in \mathbb{R}$

$\quad a < b \longleftrightarrow a + c < b + c$

(*comment*: the inverse implication ($\longleftarrow$) allows to "cancel" the number $c$, hence it represents a cancellation law.)

$T_2$ For all $a, b, c, d \in \mathbb{R}$

$$\begin{cases} a < b \\ c < d \end{cases} \longrightarrow a + c < b + d$$

## 1.3 Theorems on the product

$T_3$ **For all** $a, b, c \in \mathbb{R}$, with $c > 0$

$\quad a < b \longleftrightarrow a \cdot c < b \cdot c$

(*comment*: the inverse implication ($\longleftarrow$) allows to "cancel" the number $c$ and preserve the order.)

$T_4$ For all $a, b, c \in \mathbb{R}$ with $c < 0$

$\quad a < b \longleftrightarrow a \cdot c > b \cdot c$

(*comment*: the inverse implication ($\longleftarrow$) allows to "cancel" the number $c$ by reverting the order.)

© The Author(s), under exclusive license to Springer Nature Switzerland AG 2022
P. Toni et al., *100+1 Problems in Advanced Calculus*, Problem Books
in Mathematics, https://doi.org/10.1007/978-3-030-91863-7_1

$\mathbf{T_5}$  For all $a, b, c, d \in \mathbb{R}$ with $a, b, c, d \geq 0$

$$\begin{cases} a < b \\ c < d \end{cases} \longrightarrow a \cdot c < b \cdot d$$

$\mathbf{T_6}$  For all $a, b, c, d \in \mathbb{R}$ with $a, b, c, d \leq 0$

$$\begin{cases} a < b \\ c < d \end{cases} \longrightarrow a \cdot c > b \cdot d$$

## 1.4  Theorems on powers

$\mathbf{T_7}$  For all $a, b \in \mathbb{R}$ and $n$ odd natural number
$$a < b \longleftrightarrow a^n < b^n$$

$\mathbf{T_8}$  For all $a, b \in \mathbb{R}$, $a, b \geq 0$ and $n$ even natural number, $n \neq 0$
$$a < b \longleftrightarrow a^n < b^n$$

$\mathbf{T_9}$  For all $a, b \in \mathbb{R}$, $a, b \leq 0$ and $n$ even natural number, $n \neq 0$
$$a < b \longleftrightarrow a^n > b^n$$

$\mathbf{T_{10}}$  For all $a, x, y \in \mathbb{R}$ with $a > 1$
$$x < y \longleftrightarrow a^x < a^y$$

$\mathbf{T_{11}}$  For all $a, x, y \in \mathbb{R}$ with $0 < a < 1$
$$x < y \longleftrightarrow a^x > a^y$$

$\mathbf{T_{12}}$  For all $a, x, y \in \mathbb{R}$ with $a > 1$ and $x, y > 0$
$$x < y \longleftrightarrow \log_a x < \log_a y$$

$\mathbf{T_{13}}$  For all $a, x, y \in \mathbb{R}$ with $0 < a < 1$ and $x, y > 0$
$$x < y \longleftrightarrow \log_a x > \log_a y$$

## 1.5  Theorems on reciprocals

$\mathbf{T_{14}}$  For all $a, b \in \mathbb{R}$ with $a, b > 0$
$$a < b \longleftrightarrow \frac{1}{a} > \frac{1}{b}$$

$\mathbf{T_{15}}$  For all $a, b \in \mathbb{R}$ with $a, b < 0$
$$a < b \longleftrightarrow \frac{1}{a} > \frac{1}{b}$$

## *1.5.1  Exercise*

Assuming that $\begin{cases} -1 < a < 2 \\ 3 < b < 5 \end{cases}$ find the maximum variation permitted by these conditions to the following expressions:

(i) $3a + b^2$

(ii) $a^3 + \dfrac{1}{b}$

(iii) $\dfrac{2^a}{\log_2 b}$

## *1.5.2  Solution*

It suffices to apply the theorems on the order properties of real numbers starting from the given inequalities.

$$(i) \begin{cases} -1 < a < 2 \xleftrightarrow{T_3} -3 < 3a < 6 \xrightarrow{T_2} 6 < 3a + b^2 < 31 \\ 3 < b < 5 \xleftrightarrow{T_8} 9 < b^2 < 25 \end{cases}$$

Thus, the maximum variation permitted to $(3a + b^2)$ is $6 < 3a + b^2 < 31$.

$$(ii) \begin{cases} -1 < a < 2 \xleftrightarrow{T_7} -1 < a^3 < 8 \xrightarrow{T_2} -1 + \dfrac{1}{5} < a^3 + \dfrac{1}{b} < 8 + \dfrac{1}{3} \\ 3 < b < 5 \xleftrightarrow{T_{14}} \dfrac{1}{5} < \dfrac{1}{b} < \dfrac{1}{3} \end{cases}$$

Thus, the maximum variation permitted to $(a^3 + \frac{1}{b})$ is $-1 + \frac{1}{5} < a^3 + \frac{1}{b} < 8 + \frac{1}{3}$, that is $-\frac{4}{5} < a^3 + \frac{1}{b} < \frac{25}{3}$.

$$(iii) \begin{cases} -1 < a < 2 \xleftrightarrow{T_{10}} 2^{-1} < 2^a < 2^2 \\ 3 < b < 5 \xleftrightarrow{T_{12}} 0 < \log_2 3 < \log_2 b < \log_2 5 \\ \qquad\qquad\qquad \xleftrightarrow{T_{14}} 0 < \dfrac{1}{\log_2 5} < \dfrac{1}{\log_2 b} < \dfrac{1}{\log_2 3} \end{cases}$$

$$\xrightarrow{T_5} \dfrac{2^{-1}}{\log_2 5} < \dfrac{2^a}{\log_2 b} < \dfrac{2^2}{\log_2 3} \xrightarrow{\text{computing}} \dfrac{1}{2} \log_5 2 < \dfrac{2^a}{\log_2 b} < 4 \log_3 2$$

Thus, the maximum variation permitted to $\frac{2^a}{\log_2 b}$ is $\frac{1}{2} \log_5 2 < \frac{2^a}{\log_2 b} < 4 \log_3 2$.

# Chapter 2
# Sets, sequences, functions

## 2.1 Theoretical background

In this section, we recall the definitions of a few notions that will be used in the sequel.

### 2.1.1 Upper and lower bound of a set

**Definition 2.1** Let $\mathcal{A}$ be a subset of $\mathbb{R}$. A real number $L$ is an upper bound of $\mathcal{A}$ if and only if $x \leq L$ for all $x \in \mathcal{A}$.

**Definition 2.2** Let $\mathcal{A}$ be a subset of $\mathbb{R}$. A real number $l$ is a lower bound of $\mathcal{A}$ if and only if $x \geq l$ for all $x \in \mathcal{A}$.

### 2.1.2 Maximum and minimum of a set

**Definition 2.3** Let $\mathcal{A}$ be a subset of $\mathbb{R}$. A real number $M$ is the maximum of $\mathcal{A}$ if and only if $M$ is an upper bound of $\mathcal{A}$ and $M \in \mathcal{A}$.

**Definition 2.4** Let $\mathcal{A}$ be a subset of $\mathbb{R}$. A real number $m$ is the minimum of $A$ if and only if $m$ is a lower bound of $\mathcal{A}$ and $m \in \mathcal{A}$.

### 2.1.3 Supremum and infimum of a set

We recall two equivalent definitions of supremum of a set of real numbers.

P. Toni et al., *100+1 Problems in Advanced Calculus*, Problem Books in Mathematics, https://doi.org/10.1007/978-3-030-91863-7_2

**Definition 2.5**  The supremum of a subset $\mathcal{A}$ of $\mathbb{R}$ is the least upper bound.

**Definition 2.6**  A real number $L$ is the supremum of a subset $\mathcal{A}$ of $\mathbb{R}$ if and only if the following two conditions hold:

a) $x \leq L$ for all $x \in \mathcal{A}$ (i.e., $L$ is an upper bound);
b) for all $\epsilon > 0$ there exists $x \in \mathcal{A}$ such that $x > L - \epsilon$ (i.e., $L$ is the least upper bound).

Also for the infimum of a set of real numbers we have two equivalent definitions:

**Definition 2.7**  The infimum of a subset $\mathcal{A}$ of $\mathbb{R}$ is the greatest lower bound.

**Definition 2.8**  A real number $l$ is the infimum of a subset $\mathcal{A}$ of $\mathbb{R}$ if and only if the following two conditions hold:

a) $x \geq l$ for all $x \in \mathcal{A}$ (i.e., $l$ is a lower bound);
b) for all $\epsilon > 0$ there exists $x \in \mathcal{A}$ such that $x < l + \epsilon$ (i.e., $l$ is the greatest lower bound).

### 2.1.4  Limit point of a set

We recall two equivalent definitions of limit point.

**Definition 2.9**  The real number $c$ is a limit point of a subset $\mathcal{A}$ of $\mathbb{R}$ if and only if any neighbourhood of $c$ contains an infinite number of elements of $\mathcal{A}$.

**Definition 2.10**  The real number $c$ is a limit point of a subset $\mathcal{A}$ of $\mathbb{R}$ if and only if any neighbourhood of $c$ contains at least an element of $\mathcal{A}$ different from $c$.

*Remark 2.11*  In the definition of limit point, the condition $c \in \mathcal{A}$ is not required. Thus, there are cases where the limit point $c$ belongs to $\mathcal{A}$, and cases where it does not belong to $\mathcal{A}$.

### 2.1.5  Domain, codomain and image of a function

Given two subsets $\mathcal{A}$ and $\mathcal{B}$ of $\mathbb{R}$, and a function $f$ from $\mathcal{A}$ to $\mathcal{B}$, the set $\mathcal{A}$ is called *domain* of $f$ and the set $\mathcal{B}$ is called *codomain* of $f$. The codomain is the set of destination of $f$ and should not be confused with the *image* of $f$, defined by

$$\mathrm{Im}(f) = \{y \in \mathcal{B} : \text{there exists } x \in \mathcal{A} \text{ such that } f(x) = y\}.$$

The domain of a function $f$ is also denoted by $D(f)$.

*Remark 2.12* For real-valued functions, as those discussed in this book, if the codomain $\mathcal{B}$ is not explicitly indicated, then it is understood that $\mathcal{B} = \mathbb{R}$. We note that, for functions defined by a formula, unless otherwise indicated, it is customary to assume that the domain is the largest subset of $\mathbb{R}$ where that formula is well-defined. Thus, if we write

$$f(x) = \frac{1}{x-1}$$

without other indications, then it is understood that $\mathcal{A} = \mathbb{R} \setminus \{1\}$ and $\mathcal{B} = \mathbb{R}$. In this case the image of $f$ is

$$\mathrm{Im}(f) = \mathbb{R} \setminus \{0\}$$

which is a set strictly contained in the codomain $\mathbb{R}$. It is also customary to say that $\mathbb{R} \setminus \{1\}$ is the natural domain of definition of the function $f$ above. Note that, it is possible to consider $f$ as a function defined on a subset of $\mathbb{R} \setminus \{1\}$. For example, one may consider the function $g$ defined from $\mathcal{A} = \{x \in \mathbb{R} : x > 1\}$ to $\mathcal{B} = \mathbb{R}$ by means of the same formula above, that is

$$g(x) = \frac{1}{x-1},$$

for all $x > 1$. Note that

$$\mathrm{Im}(g) = \{y \in \mathbb{R} : y > 0\}.$$

Similarly, it is possible to extend the function $f$ by considering another function $h$ defined from $\mathcal{A} = \mathbb{R}$ to $\mathcal{B} = \mathbb{R}$ by setting, for instance,

$$h(x) = \begin{cases} \frac{1}{x-1}, & \text{if } x \neq 0, \\ 0, & \text{if } x = 0, \end{cases}$$

(here the definition of $h(0)$ is arbitrary, and one can change it depending on the circumstances). Note that

$$\mathrm{Im}(h) = \mathbb{R}.$$

Formally, functions $g$ and $h$ are different from $f$. In particular, $g$ is a restriction of $f$ and $h$ is an extension of $f$. See the next definition.

### 2.1.6  Extension and restriction of a function

**Definition 2.13** Given two functions $y = f_1(x)$ and $y = f_2(x)$ whose graphs are denoted by $G_1$, $G_2$ respectively, the function $f_2$ is an extension of $f_1$ or, equivalently, the function $f_1$ is a restriction of $f_2$, if and only if $G_1 \subseteq G_2$.

### 2.1.7  Injective, surjective and invertible functions

**Definition 2.14** Given two subsets $A$ and $B$ of $\mathbb{R}$, and a function $f$ from $A$ to $B$ we say that

(i)  $f$ is injective if $f(x_1) \neq f(x_2)$ for all $x_1, x_2 \in A$ with $x_1 \neq x_2$;

(ii)  $f$ is surjective if for every $y \in B$ there exists $x \in A$ such that $f(x) = y$;

(iii)  $f$ is invertible if $f$ is both injective and surjective. In this case, we also say that $f$ is bijective.

Moreover, if $f$ is invertible, the inverse of $f$ is the function $f^{-1}$ defined from $B$ to $A$ by setting $f^{-1}(y) = x$ where $x$ is the uniquely determined element $x \in A$ such that $f(x) = y$.

It is clear that a function is surjective if and only if $\text{Im}(f) = B$.

*Remark 2.15* If a function $f$ from $A$ to $B$ is injective then the same function $f$, considered as a function from $A$ to its image $\text{Im}(f)$, turns out to be surjective, hence invertible. Thus, it is possible to define the inverse $f^{-1}$ from $\text{Im}(f)$ to $A$. Sometimes, mostly at an elementary level, one talks of the inverse of an injective function in that wider sense. In any case, if the codomain $B$ of a function $f$ is not specified and it is tacitly assumed that $B = \mathbb{R}$, then $f$ is surjective if $\text{Im}(f) = \mathbb{R}$.

### 2.1.8  Countable and uncountable sets

**Definition 2.16** A subset $A$ of $\mathbb{R}$ is countable if it is finite or if there exists a bijection between $A$ and the set of natural numbers $\mathbb{N}$. A subset $A$ of $\mathbb{R}$ is uncountable if it is not countable.

Equivalently, a subset $A$ of $\mathbb{R}$ is countable if there exists an injective function from $A$ to $\mathbb{N}$. Besides the set of natural numbers themselves, classical examples of countable sets are the set $\mathbb{Z}$ of integer numbers and the set of rational numbers $\mathbb{Q}$. Moreover, any subset of a countable set is countable. For instance, the set of even numbers is countable since the multiplication by two defines a bijection between the set of al natural numbers and the set of even numbers. The set $\mathbb{R}$, as well as the set of all irrational numbers, is uncountable.

## 2.1.9  Algebraic and transcendental functions

**Definition 2.17** Let $f$ be a real valued function defined on an interval $I$ of $\mathbb{R}$. We say that $f$ is algebraic (over $\mathbb{R}$) if there exists a (nonzero) polynomial $P(x, y)$ in two variables $x, y$, with real coefficients, such that

$$P(x, f(x)) = 0,$$

for all $x \in I$. We say that $f$ is transcendental if it is not algebraic.

*Remark 2.18* The graph of an algebraic function is a branch of an algebraic curve. In general, an algebraic planar curve defined by a polynomial $P(x, y)$ is the set of points $(x, y) \in \mathbb{R}^2$ such that $P(x, y) = 0$. The degree of an algebraic curve is (often) defined by the degree of the polynomial $P(x, y)$, since the curve could also be identified with the equation $P(x, y) = 0$ itself (here we do not discuss the issue concerning irreducible polynomials, since we do not need it). In this book, we identify the notions of algebraic function and algebraic curve. In this sense, with a slight abuse of terminology, by degree of an algebraic function we understand the degree of the polynomial $P(x, y)$.

At an elementary level, one considers very special examples of algebraic functions obtained by a finite number of algebraic operations like summing, multiplying or dividing, and raising to fractional power. Examples of algebraic functions are:

- Polynomial functions: $y = b_0 + b_1 x + \cdots + b_p x^p$;
- Rational functions: $y = \frac{b_0 + b_1 x + \cdots + b_p x^p}{c_0 + c_1 x + \cdots + c_q x^q}$;
- Radicals: $y = \sqrt[r]{b_0 + b_1 x + \cdots + b_p x^p}$,

where all coefficients $b_i, c_i$ are real numbers and $p, q, r$ are natural numbers. Here, finding the equation $P(x, y) = 0$ satisfied by the functions $y = f(x)$ is straightforward: in the third case, it is given by $P(x, y) = y^r - b_0 - b_1 x + \cdots - b_p x^p$.

Examples of transcendental functions are:

- Trigonometric functions and their inverses : $y = \sin x$, $y = \cos x$, $y = \tan x$, $y = \arcsin x$, $y = \arccos x$, $y = \arctan x$;
- Exponential and logaritmic functions: $y = e^x$, $y = \log x$.

*Remark 2.19* One should be aware of the fact that rational functions and functions defined by radicals do not represent the whole family of algebraic functions. Indeed, the celebrated Abel-Ruffini Theorem states that an algebraic equation of degree five or larger cannot be solved by radicals in general. Thus, there exist algebraic functions which cannot be represented by using radicals. In these cases, one could talk of ultraradicals. For example, the real solution $y$ of the classical equation

$$y^5 + y - x = 0$$

**Fig. 2.1** Graph of the Bring
radical

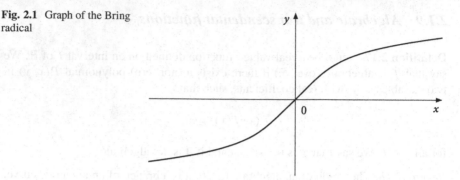

defines an algebraic function $y = f(x)$, and $f(x)$ is called ultraradical of $x$ or
Bring radical in honour of the mathematician Erland Samuel Bring (1736–1798).
See Fig. 2.1. Note that this function is nothing but the inverse of the polynomial
function $y = x^5 + x$ and is increasing.

We believe that keeping in mind the above classification of functions is quite
useful in order to be aware of the level of complexity of a problem, even if one
forgets ultraradicals and focuses the attention on elementary algebraic functions
such as those listed above and all others commonly discussed in standard textbooks.

## 2.2  Problems

### 2.2.1  Problem 1

Give an example of a countable set which is bounded from above and below and
which does not have a minimum.

### 2.2.2  Problem 2

Give an example of a countable set $\mathcal{A}$ which is bounded from above and below and
which is strictly contained in the set $\mathcal{B}$ of its limit points.

### 2.2.3  Problem 3

Give an example of a countable set $\mathcal{A}$ with a limit point which is different both from
the supremum of $\mathcal{A}$ and the infimum of $\mathcal{A}$.

### 2.2.4 Problem 4

Discuss the existence of an upper bound for the sequence

$$x_n = \frac{n^2 + 1}{n + 1}, \quad n \in \mathbb{N}.$$

### 2.2.5 Problem 5

Find the supremum of the following sequence

$$x_n = \frac{2n + 1}{n + 3}, \quad n \in \mathbb{N}.$$

### 2.2.6 Problem 6

By using the definition, say whether the infimum of the sequence

$$x_n = \frac{3n + 1}{n}, \quad n \in \mathbb{N}_0,$$

is equal to 2.

### 2.2.7 Problem 7

Find the infimum and supremum of the sequence

$$x_n = \frac{n - 1}{n^2 + 1}, \quad n \in \mathbb{N}.$$

### 2.2.8 Problem 8

Find the infimum and supremum of the sequence

$$x_n = \begin{cases} \frac{n}{4 - n^2}, & \text{if } n \in \mathbb{N}, \ n \neq 2, \\ 1, & \text{if } n = 2. \end{cases}$$

### *2.2.9  Problem 9*

Find, if any, the infimum and supremum of the sequence defined by the following recurrence formula

$$\begin{cases} a_{n+1} = \sqrt{a_n}, \\ a_0 > 0. \end{cases}$$

### *2.2.10  Problem 10*

Consider Fig. 2.2 and answer the following questions:

(a)  Is parabola $\mathcal{P}_1$ the graph of a function?
(b)  Is parabola $\mathcal{P}_2$ the graph of a function?

**Fig. 2.2** Problem 10

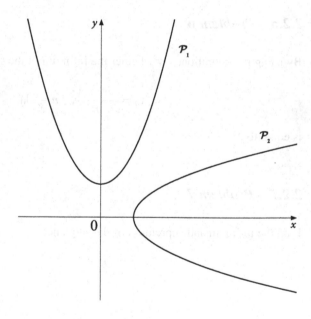

### *2.2.11  Problem 11*

Which one of the following functions defines the same function $y = x$?

(1)  $y = \sqrt{x^2}$
(2)  $y = \sqrt[3]{x^3}$
(3)  $y = (\sqrt{x})^2$

(4) $y = (\sqrt{|x|})^2$
(5) $y = \log_e e^x$
(6) $y = e^{\log_e x}$

## 2.2.12 Problem 12

Consider the following functions and say which ones coincide, and which ones are restrictions or extensions of others (see Fig. 2.3).

(1) $y = x$
(2) $y = \sqrt{x^2}$
(3) $y = \sqrt[3]{x^3}$
(4) $y = (\sqrt{x})^2$
(5) $y = (\sqrt{|x|})^2$

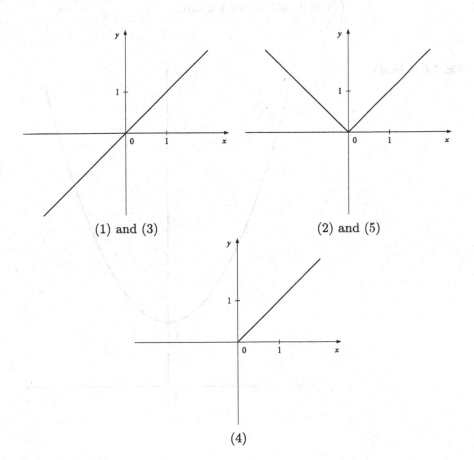

(1) and (3)    (2) and (5)

(4)

**Fig. 2.3** Problem 12

### 2.2.13 Problem 13

Give an example of a function $y = f(x)$ such that the function $\frac{1}{f(x)}$ is a proper restriction of the given function, i.e., the graph of $\frac{1}{f(x)}$ is strictly contained in the graph of $y = f(x)$.

### 2.2.14 Problem 14

Consider the function $y = x^2 + 1$ defined in the whole of $\mathbb{R}$, see Fig. 2.4. How many couples of functions $y = f_1(x)$, $y = f_2(x)$ with domains $D_1$, $D_2$ respectively, satisfy the conditions below?

$$\begin{cases} \text{both } f_1 \text{ and } f_2 \text{ are restrictions of } f, \\ \text{both } f_1 \text{ and } f_2 \text{ are injective}, \\ D_1 \cup D_2 = \mathbb{R}. \end{cases}$$

**Fig. 2.4** Problem 14

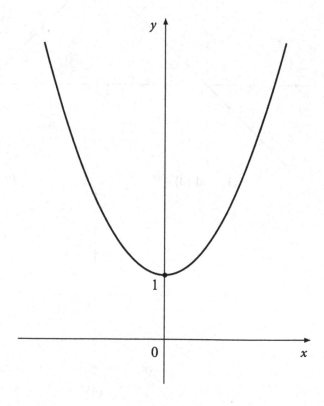

## 2.2.15 Problem 15

Recall that if the domain of a function $g$ is symmetric with respect to the origin (that is, $x \in D(g)$ if an only if $-x \in D(g)$) then $g$ is called *even* if $g(x) = g(-x)$ for all $x \in D(g)$ and is called *odd* if $g(x) = -g(-x)$ for all $x \in D(g)$.

It is well-known that any function $f$ can be written as the sum of an even function $f_E$ and an odd function $f_O$. For example:

$$f_E(x) = \frac{f(x) + f(-x)}{2},$$

$$f_O(x) = \frac{f(x) - f(-x)}{2}.$$

Is there another couple of functions $f_E$, $f_O$ as above?

## 2.2.16 Problem 16

Recall that a function $f$ defined in the whole of $\mathbb{R}$ is periodic if there exists a real number $p > 0$ such that $f(x + p) = f(x)$ for all $x \in \mathbb{R}$ (for example, $y = \sin x$). Is it true that if a non-constant function $f$ is periodic then the set of those numbers $p$ has a minimum?

## 2.3 Solutions

### 2.3.1 Solution 1

*Example*

$$x_n = \frac{1}{n}, \quad \text{for all } n \in \mathbb{N}_0.$$

See Fig. 2.5.

**Fig. 2.5** Solution 1

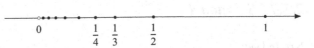

### 2.3.2   Solution 2

*Example*

$$A: 0 < x < 1, \quad \text{with } x \text{ rational},$$
$$B: 0 \le x \le 1, \quad \text{with } x \text{ rational or irrational}.$$

See Fig. 2.6.

**Fig. 2.6**  Solution 2

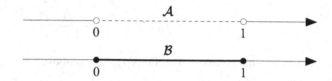

### 2.3.3   Solution 3

*Example*

$$x_n = 1 + (-1)^n \frac{1}{n}, \quad \text{for all } n \in \mathbb{N}_0.$$

See Fig. 2.7.

Limit point $= 1$
Supremum $= \frac{3}{2}$
Infimum $= 0$

**Fig. 2.7**  Solution 3

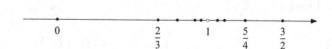

### 2.3.4   Solution 4

No. Indeed:

**1** *Intuitive method*

For $n$ large enough, we have that $n^2 + 1$ can be approximated by $n^2$, and $n + 1$ can be approximated by $n$. Thus the ratio $\frac{n^2+1}{n+1}$ can be approximated by $\frac{n^2}{n}$, that is by

$n$. Thus, for $n$ large enough, the sequence $x_n$ is essentially given by $n$, hence it is unbounded.

**2** *Formal method*

$x_n = \frac{n^2+1}{n+1} = \frac{n^2-1+2}{n+1} = \frac{(n+1)(n-1)+2}{n+1} = n - 1 + \frac{2}{n+1}$.

Since $\frac{2}{n+1} > 0$ for all $n \in \mathbb{N}$, we have that $n - 1 + \frac{2}{n+1}$ is larger than $n - 1$. Thus, since the sequence defined by $n - 1$ is unbounded, the sequence $x_n = n - 1 + \frac{2}{n+1}$ is unbounded as well.

**3** *Alternative formal method*

Assume by contradiction that there exists an upper bound $L$ for the given sequence. Then

$$x_n \leq L \quad \text{for all} \quad n \in \mathbb{N}$$

$\uparrow$ replacing the value of $x_n$
$\downarrow$

$$\frac{n^2+1}{n+1} \leq L$$

$\uparrow$ by Theorem $T_3$, since $n + 1$ is positive
$\downarrow$

$$n^2 + 1 \leq L(n+1)$$

$\uparrow$ rearranging
$\downarrow$

$$n^2 - Ln + 1 - L \leq 0 \tag{$*$}$$

The last inequality, which is given by a second degree polynomial in $n$, is satisfied only by values of $n$ belonging to a possibly empty interval $(n_1, n_2)$ of natural numbers, hence it is satisfied only by a finite number of natural numbers. This is a contradiction since we have assumed that the inequality $x_n \leq L$ holds for all $n \in \mathbb{N}$. More precisely, if $\Delta$ denotes the discriminant of the polynomial in the left-hand side of $(*)$ we have:

- If $\Delta < 0$ then $(*)$ is never satisfied;
- If $\Delta = 0$ then the polynomial in the left-hand side of $(*)$ has two coincident irrational roots, hence no roots in $\mathbb{N}$;
- If $\Delta > 0$ then inequality $(*)$ is satisfied for

$$\frac{L - \sqrt{\Delta}}{2} \leq n \leq \frac{L + \sqrt{\Delta}}{2}.$$

Thus, $(*)$ is satisfied at most for a finite number of natural numbers.

### 2.3.5   Solution 5

$\sup_n x_n = 2$. Indeed:

**1** *Intuitive method*

$$x_n = \frac{2n+1}{n+3} = \frac{(2n+6)-5}{n+3} = 2 - \frac{5}{n+3}$$

The supremum of the sequence is 2 because the ratio $\frac{5}{n+3}$ is positive and arbitrarily small as $n$ increases. Thus all numbers $x_n$ are strictly smaller than 2 and they approach 2 as much as we want.

**2** *Formal method with rigorous application of the definition* (in order to apply this method, it is necessary to known or to guess the value of the supremum in advance)

If we guess that the supremum is 2, we have to check that:

A) $\frac{2n+1}{n+3} \leq 2$ for all $n \in \mathbb{N}$ (i.e., 2 is an upper bound);
B) for all $\epsilon > 0$ there exists $n \in \mathbb{N}$ such that $\frac{2n+1}{n+3} > 2 - \epsilon$ (i.e., 2 is the least upper bound).

**PROOF OF A)**

$$\frac{2n+1}{n+3} \leq 2$$

$\Big\uparrow$ multiplying  both  sides  by  $n+3$ (Theorem $T_3$)

$$2n+1 \leq 2n+6 \quad \textit{fulfilled} \text{ for all } n \in \mathbb{N}$$

**PROOF OF B)**

$$\frac{2n+1}{n+3} > 2 - \epsilon$$

$\Big\uparrow$ multiplying  as  above  by  $n+3$ (Theorem $T_3$)

$$2n+1 > (2-\epsilon)(n+3) \quad \text{inequality of 1st degree in } n \text{ with parameter } \epsilon$$

$\Big\uparrow$ rearranging (Theorem $T_1$)

$$\epsilon n > 5 - 3\epsilon$$

⇅ multiplying both sides by $\dfrac{1}{\epsilon}$ (Theorem $T_3$)

$$n > \text{integer part of} \left( \frac{5 - 3\epsilon}{\epsilon} \right)$$

Thus, given $\epsilon$, there exists not only one natural number but also infinitely many numbers which satisfy the required inequality (as always happens in these cases). The first natural number which satisfies the condition is:

$$n_{\min} = \text{integer part of} \left( \frac{5 - 3\epsilon}{\epsilon} \right) + 1$$

**2** *Method of search of the minimal upper bound* (in order to apply this method, it is not necessary to known or to guess the value of the supremum in advance.)

If there exists an upper bound $L$ then we have:

$$x_n = \frac{2n + 1}{n + 3} \leq L \text{ for all } n \in \mathbb{N}$$

⇅ multiplying both sides by $n + 3$ (Theorem $T_3$)

$$(2 - L)n \leq 3L - 1$$

**CASE I** $\boxed{L > 2}$

$$(2 - L)n \leq 3L - 1$$

⇅ multiplying both sides by $\dfrac{1}{2 - L}$ which is negative (Theorem $T_4$)

$$n \geq \frac{3L - 1}{2 - L}$$

The last inequality is fulfilled for all $n \in \mathbb{N}$, since the ratio $\frac{3L-1}{2-L}$ is negative for all $L > 2$.

*Proof*

$$L > 2 \;\to\; 3L > 6 \;\to\; 3L - 1 > 6 - 1 \;\to\; 3L - 1 > 5 \;\to\; 3L - 1 > 0$$

Summarizing

$$L > 2 \;\to\; \begin{cases} 3L - 1 > 0 \\ 2 - L < 0 \end{cases} \;\to\; \frac{3L - 1}{2 - L} < 0$$

**CASE II** $\boxed{L < 2}$

$$(2 - L)n \le 3L - 1$$

$\Big\downarrow$ multiplying both sides by $\dfrac{1}{2 - L}$ which is positive (Theorem T$_4$)

$$n \le \frac{3L - 1}{2 - L}$$

Since the set of natural numbers is not bounded above, the last inequality is not satisfied by all natural numbers but only by those natural numbers $n$ with

$$n \le \text{integer part of } \left( \frac{3L - 1}{2 - L} \right)$$

**CASE III** $\boxed{L = 2}$

By setting $L = 2$, the initial inequality

$$(2 - L)n \le (3L - 1)$$

reads $0 \le 5$ which is *fulfilled* for all $n \in \mathbb{N}$.

*Conclusion*

The set of all upper bounds is the set of all real numbers $L$ larger or equal to 2. Thus, the least upper bound is 2.

### 2.3.6  Solution 6

We have to check that:

A) $x_n = \frac{3n+1}{n} \ge 2$ for all $n \in \mathbb{N}_0$ (i.e., 2 is a lower bound);

B) for all $\epsilon > 0$ there exists $n \in \mathbb{N}_0$ such that $\frac{3n+1}{n} < 2 + \epsilon$ (i.e., 2 is the greatest lower bound).

## PROOF OF A)

$$\frac{3n+1}{n} \geq 2$$

↑ multiplying both sides by $n$ (Theorem $T_3$)
↓

$$3n + 1 \geq 2n$$

↑ summing $-2n - 1$ in both sides (Theorem $T_1$)
↓

$$n \geq -1 \quad \textit{fulfilled} \text{ for all } n \in \mathbb{N}_0$$

## PROOF OF B)

$$\frac{3n+1}{n} < 2 + \epsilon$$

↑ multiplying as above by $n$ (Theorem $T_3$)
↓

$$3n + 1 \leq (2 + \epsilon)n \quad \text{inequality of 1st degree in } n \text{ with parameter } \epsilon > 0$$

↑ rearranging (Theorem $T_1$)
↓

$$n(1 - \epsilon) < -1$$

↑ multiplying both sides by $-1$ (Theorem $T_4$)
↓

$$n(\epsilon - 1) > 1$$

Now, we have two cases:

CASE I  $\boxed{\epsilon > 1}$

$$n(\epsilon - 1) > 1$$

↑ multiplying both sides by $\dfrac{1}{\epsilon - 1}$ which is positive (Theorem $T_3$)
↓

$$n > \frac{1}{\epsilon - 1}$$

Thus, if $\epsilon > 1$, it suffices to take a natural number $n >$ (integer part of $\frac{1}{\epsilon-1}$).

**CASE II**  $\boxed{\epsilon \le 1}$

$n(\epsilon - 1) > 1$ *never satisfied* for $n \in \mathbb{N}_0$ (since $n > 0$ and $(\epsilon - 1) \le 0$)

Thus, if $\epsilon \le 1$, there is not a natural number $n$ such that $\frac{3n+1}{n} < 2 + \epsilon$.

*Conclusion*
2 is not the infimum.

## 2.3.7   Solution 7

$\sup_{n \in \mathbb{N}}\{x_n\} = \frac{1}{5}$; $\inf_{n \in \mathbb{N}}\{x_n\} = -1$

**1** *Intuitive method*

We try to understand the behaviour of the sequence by computing a few terms:

| $x_0$ | $x_1$ | $x_2$ | $x_3$ | $x_4$ | $x_5$ | $x_6$ | $x_7$ |
|-------|-------|-------|-------|-------|-------|-------|-------|
| -1 | 0 | 1/5 | 1/5 | 3/17 | 2/13 | 5/37 | 3/25 |

The sequence is increasing from $x_0$ to $x_2 = x_3$ and decreasing from $x_3$ onwards. Thus, $x_2 = \frac{1}{5}$ is the maximum and $x_0 = -1$ is the minimum. Indeed, $x_0 = -1$ is the only negative term of the sequence and it corresponds to the only negative numerator $-1$ which is obtained for $n = 0$.

**2** *Method of study of increasing/decreasing terms*

We find those values of $n \in \mathbb{N}$ such that the sequence is increasing, that is:

$$x_n \le x_{n+1}$$

$\Big\uparrow$  replacing  the  appropriate  values  $\Big\downarrow$

$$\frac{n-1}{n^2+1} \le \frac{(n+1)-1}{(n+1)^2+1}$$

$\Big\uparrow$  multiplying both sides by $(n^2+1)[(n+1)^2+1]$ (Theorem T$_3$)  $\Big\downarrow$

$$(n-1)[(n+1)^2+1] \le n(n^2+1)$$

$\Big\uparrow$  rearranging  (Theorem T$_1$)  $\Big\downarrow$

$$n^2 - n - 2 \leq 0$$

$$\Big\uparrow \text{ solving as usual}$$

$$-1 \leq n \leq 2$$

Thus:

$$x_0 \leq x_{0+1} = x_1 \leq x_{1+1} = x_2 \leq x_{2+1} = x_3$$

Similarly, $x_n > x_{n+1}$ for all $n > 2$, that is

$$x_3 > x_{3+1} = x_4 > x_{4+1} = x_5 > x_6 > x_7 > \ldots$$

Summarising, the sequence is increasing for the first four terms and decreasing starting with the fifth term.

*Conclusion*
The minimum is $-1$ and the maximum is $\frac{1}{5}$.

## 2.3.8  *Solution 8*

$\sup_{n \in \mathbb{N}}\{x_n\} = 1; \inf_{n \in \mathbb{N}} = -\frac{3}{5}$. Indeed:

**1** *Intuitive method*

We guess the behaviour of the sequence by computing a few terms: The sequence is increasing up to $x_2$, is decreasing from $x_2$ to $x_3$, and is increasing from $x_3$ onwards, remaining negative. Thus, it is clear that the maximum equals 1 and the minimum equals $-\frac{3}{5}$.

| $x_0$ | $x_1$ | $x_2$ | $x_3$ | $x_4$ | $x_5$ | $x_6$ | $x_7$ |
|-------|-------|-------|-------|-------|-------|-------|-------|
| 0 | 1/3 | 1 | $-3/5$ | $-1/3$ | $-5/21$ | $-3/16$ | $-7/45$ |

**2** *Method of study of increasing/decreasing terms*

We study the sequence for $n \geq 3$ and we find the values of $n$ for which it is increasing, that is:

$$x_n \leq x_{n+1}$$

$$\Big\uparrow \text{ replacing the appropriate values}$$

$$\frac{n}{4-n^2} \leq \frac{n+1}{4-(n+1)^2}$$

$\uparrow$

> multiplying both sides by $(4-n^2)[4-(n+1)^2]$
> which is positive for $n \geq 3$ (Theorem $T_3$)

$\downarrow$

$$n[4-(n+1)^2] \leq (n+1)(4-n^2)$$

$\uparrow$

> rearranging

$\downarrow$

$$n^2+n+4 \geq 0 \quad \textit{fulfilled for all } n \in \mathbb{N}$$

Thus, the sequence is increasing for $n \geq 3$ and has the following behaviour:

$$x_0 = 0; \; x_1 = \frac{1}{3}; \; x_2 = 1$$

$$-\frac{3}{5} = x_3 \leq x_4 \leq x_5 \leq \cdots \leq x_n \leq \cdots \leq 0$$

*Conclusion*

We conclude that the maximum equals 1 and the minimum equals $-\frac{3}{5}$.

## 2.3.9   Solution 9

If $a_0 \leq 1$ then $\inf_n\{a_n\} = a_0$ and $\sup_n\{a_n\} = 1$.
If $a_0 \geq 1$ then $\inf_n\{a_n\} = 1$ and $\sup_n\{a_n\} = a_0$.
Indeed, by computing the first terms of the sequence we see that

$$a_0 \, ,$$

$$a_1 = \sqrt{a_0} = a_0^{\frac{1}{2}} \, ,$$

$$a_2 = \sqrt{a_1} = \sqrt{\sqrt{a_0}} = a_0^{\frac{1}{4}} \, ,$$

$$a_3 = \sqrt{a_2} = \sqrt{\sqrt{\sqrt{a_0}}} = a_0^{\frac{1}{8}} \, ,$$

etc.

It is clear that the sequence, which was defined by a recurrence relation, can be rewritten as follows:

$$a_n = a_0^{\frac{1}{2^n}}$$

**CASE** $\boxed{0 < a_0 < 1}$

We study the behaviour of the sequence:

$$n_1 \leq n_2$$

$\Big\uparrow\Big\downarrow$ Theorem $T_{10}$

$$2^{n_1} \leq 2^{n_2}$$

$\Big\uparrow\Big\downarrow$ Theorem $T_{14}$

$$\frac{1}{2^{n_1}} \geq \frac{1}{2^{n_2}}$$

$\Big\uparrow\Big\downarrow$ Theorem $T_{11}$

$$a_0^{\frac{1}{2^{n_1}}} \leq a_0^{\frac{1}{2^{n_2}}}$$

$\Big\uparrow\Big\downarrow$ substituting

$$a_{n_1} \leq a_{n_2}$$

Summarising:

$$n_1 \leq n_2 \quad \longleftrightarrow \quad a_{n_1} \leq a_{n_2}$$

which means that the sequence is increasing. Thus, the first term of the sequence is the minimum. As far as the supremum is concerned, it is useful to observe that the value of $a_n$ upon variation of $n$ is the value of the function $y = a_0^x$ upon variation of $x = \frac{1}{2^n}$. By looking at the graph of the function $y = a_0^x$ below (Fig. 2.8), we see that $0 \leq a_n \leq 1$ and $a_n$ gets closer and closer to 1 as $n$ increases, that is $1 = \sup_n\{a_n\}$. In other words,

$$\lim_{n \to \infty} a_0^{\frac{1}{2^n}} = 1.$$

**Fig. 2.8**  Solution 9

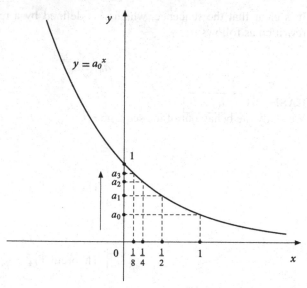

**Fig. 2.9**  Solution 10a

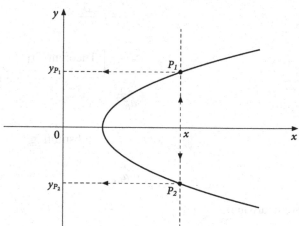

**CASE**   $\boxed{a_0 > 1}$
Similar.

## *2.3.10   Solution 10*

(a) No.

The graph of $\mathcal{P}_1$ cannot be the graph of a function in the variable $x$ because there are lines which are parallel to the $y$-axis and which intersect $\mathcal{P}_1$ at two points $P_1$ and $P_2$ (see Fig. 2.9).

(b) Yes.
Every line parallel to the $y$-axis intersects $\mathcal{P}_2$ only at one point, hence the definition of function is fulfilled (see Fig. 2.10).

**Fig. 2.10**  Solution 10b

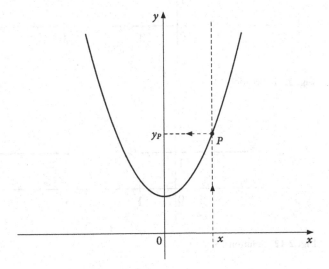

### 2.3.11   Solution 11

(2) and (5).

### 2.3.12   Solution 12

(4) is the restriction both of (1) and (3), as well as of (2) and (5). Vice-versa, both (1) and (3), as well as (2) and (5) are extensions of (4).

### 2.3.13   Solution 13

*Example*

$$f(x) = \begin{cases} 0, & \text{if } 1 < x < 3, \\ 1, & \text{if } x \leq 1 \text{ or } x \geq 3. \end{cases}$$

See Fig. 2.11.

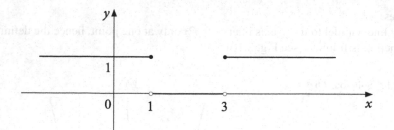

**Fig. 2.11** Solution 13

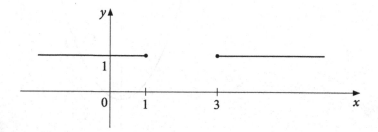

**Fig. 2.12** Solution 13

$$\frac{1}{f(x)} = \begin{cases} \text{not defined,} & \text{if } 1 < x < 3, \\ 1, & \text{if } x \leq 1 \text{ or } x \geq 3. \end{cases}$$

See Fig. 2.12.

### 2.3.14   Solution 14

Infinitely many. Indeed, the most natural couple of functions could be

$$\begin{cases} f_1(x) = x^2 + 1, \\ x < 0, \end{cases} \quad \text{and} \quad \begin{cases} f_2(x) = x^2 + 1, \\ x \geq 0, \end{cases}$$

or

$$\begin{cases} f_1(x) = x^2 + 1, \\ x \leq 0, \end{cases} \quad \text{and} \quad \begin{cases} f_2(x) = x^2 + 1, \\ x > 0. \end{cases}$$

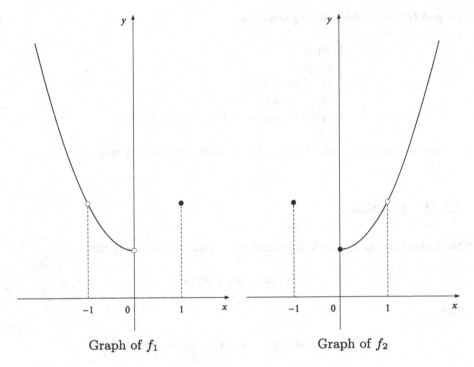

Graph of $f_1$        Graph of $f_2$

**Fig. 2.13** Solution 14

However, it is possible to define infinitely many others. For example (see Fig. 2.13):

$$f_1(x) = x^2 + 1, \quad \text{if } x < -1, \text{ or } \quad -1 < x < 0, \text{ or } x = 1,$$

$$f_2(x) = x^2 + 1, \quad \text{if } x = -1, \text{ or } \quad 0 \le x < 1, \text{ or } x > 1.$$

Generalizing the previous example, one can define

$$f_1(x) = x^2 + 1, \quad \text{if } x < -n, \text{ or } \quad -n < x < 0, \text{ or } x = n,$$

$$f_2(x) = x^2 + 1, \quad \text{if } x = -n, \text{ or } \quad 0 \le x < n, \text{ or } x > n,$$

with $n \in \mathbb{N}$.

Generalizing even more, one can define other couples of functions by observing that a couple of functions $f_1$, $f_2$ answers the question if the corresponding domains

$D_1$ and $D_2$ satisfy the following conditions

$$\begin{cases} D_1 \neq \emptyset, \\ D_2 \neq \emptyset, \\ D_1 \cap D_2 = \emptyset, \\ D_1 \cup D_2 = \mathbb{R}, \\ (x \in D_1 \ \text{and} \ x \neq 0) \leftrightarrow -x \in D_2. \end{cases}$$

Note that the partitions $\{D_1, D_2\}$ of $\mathbb{R}$ as above are infinitely many.

## 2.3.15   Solution 15

No. Indeed, if $\varphi$ is an even function and $\psi$ is an odd function such that

$$f(x) = \varphi(x) + \psi(x)$$

then

$$f(-x) = \varphi(-x) + \psi(-x) = \varphi(x) - \psi(x)$$

hence

$$\frac{f(x) + f(-x)}{2} = \varphi(x)$$

and

$$\frac{f(x) - f(-x)}{2} = \psi(x).$$

## 2.3.16   Solution 16

No.

*Example*

$$f(x) = \begin{cases} 1, & \text{if } x \text{ is rational}, \\ 0, & \text{if } x \text{ is irrational}. \end{cases}$$

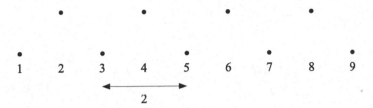

**Fig. 2.14** Solution 16

If $r$ is any fixed rational number, we have that

- If $x$ is rational then $x + r$ is rational, hence $f(x + r) = f(x) = 1$;
- If $x$ is irrational then $x + r$ is irrational, hence $f(x + r) = f(x) = 0$.

Thus, $f(x + r) = f(x)$ for all $x \in \mathbb{R}$ and clearly the set $\{r \in \mathbb{Q} : r > 0\}$ has no minimum.

*Remark* Maybe the intuitive idea of a minimum period relies on a Pythagorean image of point for which the minimum period is 2 (see Fig. 2.14). These are the tricks played by the notion of point without dimensions proposed by Euclid.

# Chapter 3
# Limits of functions, continuity

## 3.1 Theoretical background

In this section, we recall the definitions of a few notions that will be used in the sequel.

### 3.1.1 Limits of functions – finite case

**Definition 3.1 (Classical definition)** Let $\mathcal{A}$ be a subset of $\mathbb{R}$ and $f$ a function from $\mathcal{A}$ to $\mathbb{R}$. Assume that $c \in \mathbb{R}$ is a limit point of $\mathcal{A}$. We say that a real number $l$ is the limit of $f$ as $x$ tends to $c$, and we write

$$\lim_{x \to c} f(x) = l,$$

if for all $\epsilon > 0$ there exists $\delta > 0$ such that

$$\begin{cases} |x - c| < \delta \\ x \in \mathcal{A}, \ x \neq c \end{cases} \rightarrow |f(x) - l| < \epsilon$$

We note that the conditions $|x - c| < \delta$ and $|f(x) - l| < \epsilon$ can be written in the equivalent form $x \in I_c(\delta)$ and $f(x) \in I_l(\epsilon)$ where $I_c(\delta) = \{x \in \mathbb{R} : \ |x - c| < \delta\}$ and $I_l(\epsilon) = \{x \in \mathbb{R} : \ |x - l| < \epsilon\}$. In general, for any $c \in \mathbb{R}$ and $r > 0$, one may set

$$I_c(r) = \{x \in \mathbb{R} : \ |x - a| < r\}$$

and call it *(circular) neighborhood* of center $c$ and radius $r$.

© The Author(s), under exclusive license to Springer Nature Switzerland AG 2022
P. Toni et al., *100+1 Problems in Advanced Calculus*, Problem Books
in Mathematics, https://doi.org/10.1007/978-3-030-91863-7_3

In order to verify that $\lim_{x \to c} f(x) = l$, typically one has to check that the set of all $x \in \mathcal{A}$ such that $|f(x) - l| < \epsilon$ contains the set $(\mathcal{A} \cap I_c(\delta)) \setminus \{c\}$ for some $\delta > 0$. In many concrete examples and problems, one often finds that the condition $|f(x) - l| < \epsilon$ is satisfied for all points $x$ belonging to an interval $(\alpha, \beta)$ with $\alpha < c < \beta$. This interval is not necessarily a circular neigborhood of $c$, but certainly contains circular neigborhoods of $c$ as required in the classical definition of limit. This suggests to consider not only circular neighborhoods as above but also asymmetric neighborhoods of the form

$$I_c = (\alpha, \beta), \tag{3.1}$$

for all $\alpha, \beta \in \mathbb{R}$, with $\alpha < c < \beta$. This leads to the following modern definition of limit which is stated using more general neighborhoods[1] as in (3.1), possibly asymmetric. This definition is obviously equivalent to the classical one.

**Definition 3.2 (Modern definition)** Let $\mathcal{A}$ be a subset of $\mathbb{R}$ and $f$ a function from $\mathcal{A}$ to $\mathbb{R}$. Assume that $c \in \mathbb{R}$ is a limit point of $\mathcal{A}$. We say that a real number $l$ is the limit of $f$ as $x$ tends to $c$, and we write

$$\lim_{x \to c} f(x) = l,$$

if for any neighborhood $I_l$ of $l$ there exists a neighborhood of $c$ such that

$$\begin{cases} x \in I_l \\ x \in \mathcal{A}, \ x \neq c \end{cases} \rightarrow f(x) \in I_l$$

The modern definition of limit can be used, as it is, to consider not only finite limits as above, but also limits at infinity and infinite limits. Namely, one can consider the cases where $c, l \in \{+\infty, -\infty, \infty\}$ provided one clarifies what is meant by neighborhood of $+\infty$, $-\infty$, and $\infty$. These notions strongly depends on different images of the real line as discussed below.

---

[1] According to the contemporary standard definition, a neighborhood of a real number $c$ in the Euclidean topology is any set of real numbers containing an open interval which contains $c$. Moreover, the family of circular neighborhoods $I_c(\delta)$ of $c$ represents what is called a 'basis of neighborhoods' of $c$, i.e., a family of neighborhoods such that any arbitrary neighborhood contains a neighborhood of the chosen family. In the definition of limit, one can use any basis of neighborhoods. The use of circular neighborhoods is one option, the classical one. As we shall see in concrete examples, sometimes the use of asymmetric neighborhoods of the type (3.1) reduces the complexity of the calculations involved in certain problems.

## 3.1.2 Image of the real line and neighborhoods at infinity

In order to study the limits of functions of one real variable it is convenient to introduce the symbols $+\infty$, $-\infty$ or $\infty$ beside the set of real numbers. By doing so, the intuitive image of the real line changes a lot.

In the beginning, the real line is understood as a set of points which is unbounded in both directions (see Fig. 3.1).

Then, the real line is endowed with the two extreme points $+\infty$ and $-\infty$, which cannot be "overstepped": $+\infty$ is the largest and $-\infty$ is the smallest (see Fig. 3.2).

Having this image in mind, it is natural to say that a neighborhood $I_{+\infty}$ of $+\infty$ is any set of the form $(M, +\infty)$ and a neighborhood $I_{-\infty}$ of $-\infty$ is any set of the form $(-\infty, M)$. In both cases, $M$ can be any real number.

Finally, by introducing $\infty$ (instead of $+\infty$ and $-\infty$), the real line assumes a circular character where the two extremes coincide (see Fig. 3.3).

Having this image in mind, it is natural to say that a neighborhood $I_{\infty}$ of $\infty$ is any set of the form $(-\infty, M) \cup (N, +\infty)$ with $M, N$ real numbers.

In fact, one advantage of the circular representation consists in the fact that it allows to have the same geometric image of the neighbourhood of a point both in the finite and in the infinite case. For example, as one can see in Fig. 3.4, we can represent in the same way a neighbourhood $I_5$ of 5, a neighbourhood $I_{-3}$ of $-3$ and a neighbourhood $I_{\infty}$ of $\infty$.

---

**Fig. 3.1** Line unbounded in both directions

$-\infty$ • ———————————————————————— • $+\infty$

**Fig. 3.2** Line with extremes

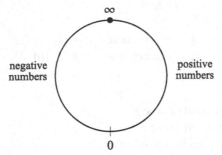

**Fig. 3.3** Circular representation of the real line

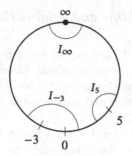

**Fig. 3.4** Unitary image of neighborhoods

The differences are only in the algebraic representations. For example, the neighbourhoods above can be represented as follows

$$I_5 = (4, 7) \qquad\qquad 4 < x < 7$$

values inside the interval $(4, 7)$

$$I_\infty = (-\infty, 2) \cup (10, +\infty) \quad x < 2 \text{ or } x > 10$$

values outside the interval $[2, 10]$

From an algebraic point of view, we may say that:

| |
|---|
| Neighbourhood of a finite value is synonymous with *internal values* |

| |
|---|
| Neighbourhood of $\infty$ is synonymous with *external values* |

Thus:

$$x \in I_5 \quad \text{becomes} \quad x \in (4, 7)$$
$$x \in I_\infty \quad \text{becomes} \quad x \notin [2, 10]$$

Questions:

- $x < 3$ is a neighbourhood of $-\infty$?
- $x > 5$ is a neighbourhood of $+\infty$?
- $x < 2$ or $x > 6$ is a neighbourhood of $\infty$?

The answer is positive in all the three cases.

Summing up, the neighbourhoods of $-\infty$, $+\infty$, $\infty$ can be algebraically represented as follows:

$I_{-\infty}:$  $x < \alpha$
$I_{+\infty}:$  $x > \beta$
$I_{\infty}:$  $x \notin [\alpha, \beta]$

where $\alpha$, $\beta$ are arbitrary real numbers.

*Remark 3.3* In some books, it is often written:

$I_{-\infty}:$  $x < -N$
$I_{+\infty}:$  $x > N$
$I_{\infty}:$  $|x| > N$ which is equivalent to $x \notin [-N, N]$

where $N > 0$. With regard to this habit, we observe that the assumption $N > 0$ is not essential but is useful just to simplify computations.

### 3.1.3 Limits of functions – general case (new symbols)

The discussion above, suggests that in order to embrace all cases in the definition of limit, one could use the general notation

$$\lim_{x \to \square} f(x) = \triangle$$

where the "metasymbols" $\square$ and $\triangle$ represent one of the following symbols:

$\square$  stands for $c, c^+, c^-, \infty, +\infty, -\infty$
$\triangle$  stands for $l, l^+, l^-, \infty, +\infty, -\infty$

Note that the cases $c^+$ and $c^-$ represent the classical cases of right and left limits respectively, and $l^+$, $l^-$ represent the convergence to $l$ from above and below respectively.

Moreover, it is convenient to use the symbol $I_{\square}$ to denote a neighbourhood of $\square$, whatever $\square$ is. In particular, if $\square = c^+$ (respectively, $\square = c^-$) then $I_{\square}$ will be a right (respectively, left) neighbourhood of $c$. Recall that right neigborhoods of $c$ are the intervals of the form

$$I_{c^+} = [c, \beta)$$

for all $\beta > c$, and left neigborhoods of $c$ are the intervals of the form

$$I_{c^-} = (\alpha, c]$$

for all $\alpha < c$.

We are now ready to give the following general definition which is deliberately written in a formal way also for mnemonic purposes.

**Definition 3.4 (Modern definition – general case)** Let $\square, \triangle \in \{c, c^+, c^-, \infty, +\infty, -\infty\}$. We say that

$$\lim_{x \to \square} f(x) = \triangle$$

if for any neighborhood $I_\triangle$ there exists a neighborhood $I_\square$ such that

$$\begin{cases} x \in I_\square \\ x \in D(f), \ x \neq \square \end{cases} \to f(x) \in I_\triangle.$$

In the previous definition it is understood that $\square$ must be a limit point of the domain $D(f)$ of the function $f$, that is, there must be infinitely many elements of $D(f)$ arbitrarily close to $\square$ (from the right in the case of right limits and from left in the case of left limits).

The reader may find it useful to write the previous definition for specific values of $\square$ and $\triangle$. For instance, if $\square = \triangle = +\infty$, according to the previous definition we have that

$$\lim_{x \to +\infty} f(x) = +\infty$$

if and only if for any $M \in \mathbb{R}$ there exists $N \in \mathbb{R}$ such that

$$\begin{cases} x > N \\ x \in D(f) \end{cases} \to f(x) > M$$

The next table describes in a schematic way the various definitions of limits and includes a representation of the logical scheme.

## 3.2   Definition of limit

$$\lim_{x \to \square} f(x) = \triangle \text{ with } \square \in \{c, c^+, c^-, \infty, +\infty, -\infty\}$$
$$\triangle \in \{l, l^+, l^-, \infty, +\infty, -\infty\}$$

| Logical scheme | Modern definition (neighborhood-neighborhood) | "Semi-modern definition" (ray-neighborhood) special case $\lim_{x\to\square} f(x) = l$ | Classical definition (ray-ray) special case $\lim_{x\to c} f(x) = l$ |
|---|---|---|---|
| For any move of the *black* there exists a response of the *white* which delivers<br><br>*checkmate* | For all $I_\Delta$ there exists $I_\square$ such that $\begin{cases} x \in I_\square \\ x \neq \square \\ x \in D(f) \end{cases} \to f(x) \in I_\Delta$ | For all $\epsilon > 0$ there exists $I_\square$ such that $\begin{cases} x \in I_\square \\ x \neq \square \\ x \in D(f) \end{cases} \to |f(x) - l| < \epsilon$ | For all $\epsilon > 0$ there exists $\delta > 0$ such that $\begin{cases} |x - c| < \delta \\ x \neq c \end{cases} \to |f(x) - l| < \epsilon$ |

*Remark* As far as the definition of limit is concerned, we note that the condition $x \in D(f) = $ domain of $f$ is often omitted in the high-school textbooks. That condition is necessary for peculiar and very significant functions.

Note also that the condition $x \neq \square$ is used only in the finite case $x \neq c$ since usually one doesn't write $x \neq \infty$.

### 3.2.1   Continuity

We recall here the definition of continuous functions:

**Definition 3.5** Let $\mathcal{A}$ be a subset of $\mathbb{R}$ and $f$ a function from $\mathcal{A}$ to $\mathbb{R}$. Assume that $c \in \mathcal{A}$ is a limit point of $\mathcal{A}$. We say that $f$ is continuous at $c$ if

$$\lim_{x \to c} f(x) = f(c). \tag{3.2}$$

We also say that $f$ is continuous if it is continuous at any limit point of $\mathcal{A}$ belonging to $\mathcal{A}$.

It is a matter of folklore to say that a continuous function is a function the graph of which can be drawn without lifting the chalk from the blackboard. This vision is not completely wrong if one considers continuous functions defined on an interval. However, as it will be clear from the problems below, one has to be very cautious, also because the notions of limit and continuity do not require that the domain of a function is an interval or the union of intervals. In particular, we note that the notion of continuity makes sense only at those points where $f$ is defined. For example, for a function defined by the formula

$$f(x) = \frac{\sin x}{x}, \tag{3.3}$$

the natural domain of definition is $D(f) = \mathbb{R} \setminus \{0\}$. Then one is tempted (and some textbooks do it) to say that this function is discontinuous at the point $x = 0$ because one has to lift the chalk at $x = 0$ when plotting its graph at the blackboard. However, the same function can be extended by continuity at $x = 0$ by setting $f(0) = 1$, in which case one would obtain a continuous function because condition (3.2) would be fulfilled. This example shows that it would be better to talk about continuity and discontinuity only for those limit points of the domain where the function is defined, otherwise one could get contradictory notions. See the next section.

### 3.2.2  *Classification of discontinuities*

Students may find it useful to classify the discontinuities of a function according to its behaviour around the limit points under consideration. Here we recall the following classification (note that other classifications are used in the literature).

**Definition 3.6** Let $\mathcal{A}$ be a subset of $\mathbb{R}$ and $f$ a function from $\mathcal{A}$ to $\mathbb{R}$. Assume that $c \in \mathcal{A}$ is a limit point of $\mathcal{A}$ and that $f$ is discontinuous at $c$. We say that

(i)  the discontinuity of $f$ at $c$ is of the first kind if both the right and left limits of $f$ at $c$ exist, are finite and

$$\lim_{x \to c^-} f(x) \neq \lim_{x \to c^+} f(x).$$

In this case, we also say that $f$ has a jump discontinuity at $c$;

(ii)  the discontinuity of $f$ at $c$ is of the second kind if it is not of the first kind and the limit of $f$ at $c$ does not exist in $\mathbb{R}$;

(iii)  the discontinuity of $f$ at $c$ is of the third kind if the limit of $f$ at $c$ exists, is finite and

$$\lim_{x \to c} f(x) \neq f(c).$$

In this case, we also say that $f$ has a removable discontinuity at $c$.

Obviously, case (i) is of interest only when it makes sense to consider both left and right limits, that is, when there are infinitely many elements of $\mathcal{A}$ arbitrarily close to $c$ both on the left-hand side and on the right-hand side of $c$.

We now give a few elementary examples. The function $f$ from $\mathbb{R}$ to $\mathbb{R}$ defined by

$$f(x) = \begin{cases} 1, & \text{if } x \geq 0, \\ -1, & \text{if } x < 0, \end{cases}$$

has a discontinuity of the first kind at $x = 0$. The function $g$ from $\mathbb{R}$ to $\mathbb{R}$ defined by

$$g(x) = \begin{cases} \sin \frac{1}{x}, & \text{if } x \neq 0, \\ 0, & \text{if } x = 0 \end{cases}$$

has a discontinuity of the second kind at $x = 0$. The function $h$ from $\mathbb{R}$ to $\mathbb{R}$ defined by

$$h(x) = \begin{cases} 0, & \text{if } x \neq 0, \\ 1, & \text{if } x = 0 \end{cases}$$

has a discontinuity of the third kind at $x = 0$: this discontinuity is called removable because if one re-defines $h$ at zero by setting $h(0) = 0$, the new function would be continuous.

Note that also the function $k$ from $\mathbb{R}$ to $\mathbb{R}$ defined by

$$k(x) = \begin{cases} \log |x|, & \text{if } x \neq 0, \\ 0, & \text{if } x = 0 \end{cases}$$

has a discontinuity of the second kind at $x = 0$ because the

$$\lim_{x \to 0} k(x) = -\infty$$

but $-\infty$ is not a real number.

It is tempting to extend the classification of the points of discontinuity of a function in order to include the case of limit points of the domain which do not belong to the domain itself. This can be easily done by removing in the classification above the condition that the point $c$ belongs to the domain $\mathcal{A}$ of the function $f$. One is obviously free to do it and some textbooks do it. In this way, according to this extended notion of discontinuity, one would have that the following functions $\varphi, \psi, \eta$ defined on $\mathbb{R} \setminus \{0\}$ by the formulas

$$\varphi(x) = \frac{x}{|x|}, \quad \psi(x) = \frac{1}{x}, \quad \eta(x) = \frac{1 - \cos x}{x^2}$$

for all $x \in \mathbb{R} \setminus \{0\}$, are discontinuous at $x = 0$. In particular, $\varphi, \psi$, and $\eta$ would have a discontinuity of the first, the second and the third kind respectively.

On one hand, this would be convenient, because the natural domain of definition of these functions is $\mathbb{R} \setminus \{0\}$. On the other hand, as we have already mentioned, one should be aware of the fact that, having this extended notion of discontinuity could lead to some contradictory situations, at least at the level of terminology. For example, the function $\eta$ can be naturally extended by continuity to the whole of $\mathbb{R}$ by setting

$$\eta(0) = \frac{1}{2}.$$

In other words, the function $\tilde{\eta}$ defined on the whole of $\mathbb{R}$ by setting

$$\tilde{\eta}(x) = \begin{cases} \frac{1 - \cos x}{x^2}, & \text{if } x \neq 0, \\ \frac{1}{2}, & \text{if } x = 0 \end{cases}$$

is continuous. It is clearly embarrassing to claim that $\tilde{\eta}$ is continuous while its restriction $\eta$ is discontinuous! One may think that this issue is not particularly important, since this example concerns a *removable* discontinuity. However, the same problem would occur also with some discontinuities of the second kind. For example, having in mind the extended real line $\mathbb{R} \cup \{\infty\}$ described in Fig. 3.3, one may consider the function $\tilde{\psi}$ from $\mathbb{R}$ to $\mathbb{R} \cup \{\infty\}$ defined by

$$\tilde{\psi}(x) = \begin{cases} \frac{1}{x}, & \text{if } x \neq 0, \\ \infty, & \text{if } x = 0. \end{cases}$$

Note that here the codomain of $\tilde{\psi}$ is not $\mathbb{R}$, hence, strictly speaking, our definition of continuity does apply to $\tilde{\psi}$. However, it would be natural to extend the definition of continuity also to the case of functions assuming infinite values.[2] In this sense, since $\lim_{x \to 0} \tilde{\psi}(x) = \tilde{\psi}(0) = \infty$, the function $\tilde{\psi}$ would be continuous as a function from $\mathbb{R}$ to $\mathbb{R} \cup \{\infty\}$. Again, it would be a bit weird to have a continuous function like $\tilde{\psi}$ and to claim that its restriction $\psi$ is discontinuous! Note that in this book, we shall not consider functions with infinite values, and function $\tilde{\psi}$ is used here only to discourage the use of a misleading terminology.

In conclusion, for formulas like those defining functions $\varphi$, $\psi$, $\eta$, as well as for formula (3.3) and others, it would be preferable to talk about *singularity*[3] rather than discontinuity at the point $x = 0$. However, we shall not discuss this issue further.

### 3.2.3  *Geometric interpretation of the definition of the limit of a function*

Consider a real valued function $f$ defined on $\mathbb{R}$ and the limit

$$\lim_{x \to c} f(x) = l,$$

where $c$ and $l$ are real numbers.

Students may find the definition of that limit a bit complicated and artificial, running the risk of repeating it in a ritual way without an appropriate comprehension. On the other hand the notion of function involves the couple domain-codomain, hence the difficult interplay between neighbourhoods is inevitable. However, passing from a *function* to its *graph*, or equivalently, passing from the $x$ and $y$-axes to the Cartesian plane, allows a unitary view of the 'development' of the graph in the plane. This is possible by passing from the couple of neighbourhoods defined on each axis to rectangles defined in the Cartesian plane. The family of couples of neighbourhoods is then replaced by a family of rectangles, nested, one inside another, like Chinese boxes. If this family of rectangles would have a three-

---

[2] This is standard in the study of Mathematical Analysis in metric spaces and more general topological spaces.

[3] This is done for example in the theory of holomorphic functions.

dimensional displacement, then the definition of limit of a function would produce a tunnel, a funnel, a bellow as in certain folding cameras, into which the graph squeezes and eventually comes out of the limit $(c, l)$, a point which belongs to all rectangles.

Indeed, consider a generic point $P = (x, f(x))$ of the graph of $f$. Reading the definition of limit step by step, we have:

$$f(x) \in I_l \qquad \longleftrightarrow \qquad P \in S_1 \quad \text{(horizontal strip)}$$

See Fig. 3.5.

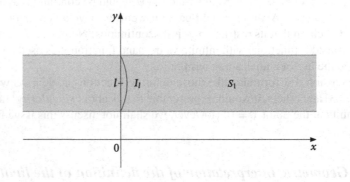

**Fig. 3.5**  Horizontal strip

$$x \in I_c \qquad \longleftrightarrow \qquad P \in S_2 \quad \text{(vertical strip)}$$

See Fig. 3.6.

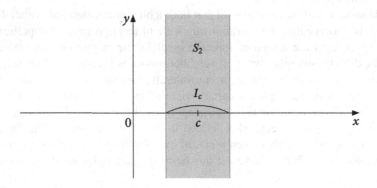

**Fig. 3.6**  Vertical strip

$$(x \in I_c \quad \to \quad f(x) \in I_l) \quad \longrightarrow \quad P \in R = S_1 \cap S_2 \quad \text{(rectangle)}$$

See Fig. 3.7.

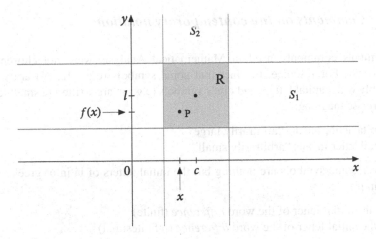

**Fig. 3.7** Rectangle

Thus, all points of the graph contained in $S_2$ belong to the rectangle $R$. Taking another neighbourhood $I_l'$ with $I_l' \subset I_l$, we get a new rectangle $R'$ with $R' \subset R$ and so on (see Fig. 3.8).

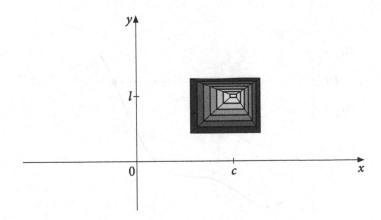

**Fig. 3.8** The tunnel

Starting now from the very synthetic image of the tunnel and going back to the definition which has been used to construct it, we realise that the definition of limit of a function is made of a minimal number of conditions which are compatible with the Cartesian representation of the function. Thus, the complexity of the definition is not reducible.

### 3.2.4   Comments on the contemporary notation

The symbols commonly used in Mathematical Analysis were not chosen in a random way. For instance, the fact that some symbols (e.g., $M$, $N$) are written preferably with capital letters and other symbols (e.g., $\epsilon$) are written in small letters, has a precise meaning:

a capital letter means "arbitrarily large"
a small letter means "arbitrarily small"

Moreover, some symbols are nothing but the initial letters of latin or greek words. For example:

$\Delta$ is the initial letter of the word *difference* (finite)
$\delta$ is the initial letter of the word *difference* (infinitesimal)

Finally, in the standard definition of the limit of a function, the letters $\epsilon$, $\delta$ are often used in tandem. This is due to the fact that in the Greek alphabet the letter $\epsilon$ immediately follows the letter $\delta$, as in the Latin alphabet the letter $y$ (which is used for the $y$-axis) follows the letter $x$ (which is used for the $x$-axis). See Fig. 3.9.

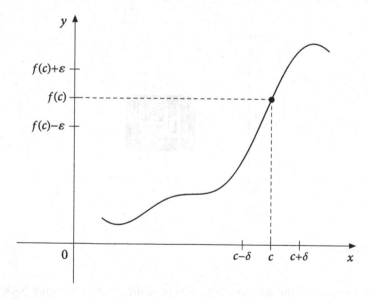

**Fig. 3.9** The alphabetic order

## 3.2.5 Continuity of the function $y = \sin x$

**Step 1** We prove that $\lim_{x \to 0^+} \sin x = 0$

To do so, we can assume that $0 < x < \pi/2$. We consider the trigonometric circle and an angle $\widehat{POQ}$ the measure of which is $x$ radiants (see Fig. 3.10). This means that

$$x = \frac{\widehat{PQ}}{\text{ray}} = \frac{\widehat{PQ}}{1} = \widehat{PQ}.$$

We consider now the point $P'$, symmetric of $P$ with respect to the $x$-axis, and we draw the chord $PP'$ which intersects the segment $OQ$ at the point $H$. We note that $\overline{PH} = \sin x$. As is known from elementary geometry, we have that

$$0 \le \overline{PP'} \le \widehat{PP'} \quad \text{by the circle postulate } chord < arc,$$

hence, dividing by 2 we get

$$0 \le \overline{PH} \le \widehat{PQ}.$$

By substituting the corresponding values, we have

$$0 \le \sin x \le x. \tag{3.4}$$

**Fig. 3.10** Step 1

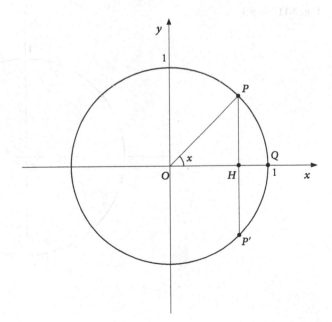

By passing to the limit in (3.4) as $x \to 0^+$ and using the Squeeze Theorem, we get

$$\lim_{x \to 0^+} \sin x = 0.$$

**Step 2** We prove that $\lim_{x \to 0^-} \sin x = 0$.
To do so, we can directly assume that $-\pi/2 < x < 0$. This implies that $\pi/2 > -x > 0$ and by what we have proved in Step 1, we have that $0 \le \sin(-x) \le -x$, hence $0 \le -\sin x \le -x$ and

$$0 \ge \sin x \ge x. \tag{3.5}$$

By passing to the limit in (3.5) as $x \to 0^-$ and using the Squeeze Theorem, we get

$$\lim_{x \to 0^-} \sin x = 0.$$

Summing up, we have

$$\lim_{x \to 0} \sin x = \lim_{x \to 0^+} \sin x = \lim_{x \to 0^-} \sin x = 0.$$

**Step 3** We prove that $\lim_{x \to 0} \cos x = 1$.
Similarly to what we we have done in Step 2, we consider in the trigonometric circle an angle $\widehat{POQ} = x$ and the triangle $POH$ as in Fig. 3.11.

**Fig. 3.11** Step 3

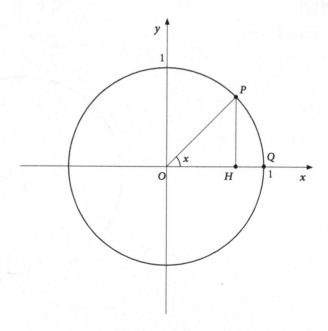

By recalling that in every triangle, the length of each side is greater than the difference of the lengths of the other two sides, we have

$$\overline{OP} - \overline{PH} \leq \overline{OH} \leq 1$$

and replacing the corresponding values

$$1 - \sin x \leq \cos x \leq 1. \tag{3.6}$$

Thus, since $\lim_{x \to 0^+} (1 - \sin x) = 1$, by the Squeeze Theorem and inequality (3.6), we get $\lim_{x \to 0^+} \cos x = 1$. By observing that $\cos x$ is an even function, we can conclude that

$$\lim_{x \to 0} \cos x = \lim_{x \to 0^+} \cos x = \lim_{x \to 0^-} \cos x = 1.$$

**Step 4** Finally, we prove the continuity of the sin function, i.e., $\lim_{x \to c} \sin x = \sin c$. To do so, we set $x - c = t$ and we note that $t \to 0$ as $x \to c$. We can represent this substitution as follows:

$$x - c = t \longleftrightarrow x = t + c$$
$$x \to c \quad \longleftrightarrow t \to 0$$

Thus, we have

$$\lim_{x \to c} \sin x = \lim_{t \to 0} \sin(t + c) \overset{(A)}{=} \lim_{t \to 0} (\sin t \cos c + \cos t \sin c)$$

$$\overset{(B)}{=} \lim_{t \to 0} \sin t \cos c + \lim_{t \to 0} \cos t \sin c$$

$$\overset{(C)}{=} \lim_{t \to 0} \sin t \lim_{t \to 0} \cos c + \lim_{t \to 0} \cos t \lim_{t \to 0} \sin c$$

$$\overset{(D)}{=} 0 \cdot \cos c + 1 \cdot \sin c = \sin c$$

(A) Trigonometric identity;
(B) Theorem of the limit of the sum of functions;
(C) Theorem of the limit of the product of functions;
(D) continuity at zero of the functions $y = \sin x$ and $y = \cos x$, and continuity of the constant function.

## 3.3   Problems

### 3.3.1   Problem 17

Using the definition of the limit of a function, prove that

(1) $\lim\limits_{x \to 2} \sqrt{25(x^2 - 3)} = 5$

(2) $\lim\limits_{x \to 1} \sqrt{\log_{\frac{1}{2}} \dfrac{1}{8x + 8}} = 2$

(3) $\lim\limits_{x \to 5} 9^{\sqrt{\frac{x-5}{x^2 - 6x + 5}}} = 3$

(4) $\lim\limits_{x \to 1^+} 2^{|\log_x 3|} = +\infty$

(5) $\lim\limits_{x \to +\infty} \sqrt[3]{\log_2 \sin \dfrac{1}{x}} = -\infty$

### 3.3.2   Problem 18

Is it true that if $\lim_{x \to c} f(x) = l$ then $l$ is a limit point of the image of the function $f$?

### 3.3.3   Problem 19

Is it true that if $\lim_{x \to c} f(x) = l$ and $f$ is not constant in any neighbourhood of $c$ then $l$ is a limit point of the image of the function $f$?

### 3.3.4   Problem 20

Is it true that if $\lim_{x \to c} f(x) = l$ and $l$ is a limit point of the image of the function $f$ then $f$ is not constant in any neighbourhood of $c$?

### 3.3.5   Problem 21

Assume that $l$ is a limit point of the image of the function. Is there a real number $c$ such that $\lim_{x \to c} f(x) = l$?

### 3.3.6 Problem 22

Assume that $l$ is the unique limit point of the image of a function $f$. Is there a real number $c$ such that $\lim_{x \to c} f(x) = l$?

### 3.3.7 Problem 23

Recall that the theorem on uniqueness of limits states that if $\lim_{x \to c} f(x) = l_1$ and $\lim_{x \to c} f(x) = l_2$ then $l_1 = l_2$. We now 'prove' that this theorem is *false* by giving a counterexample. Consider the function

$$y = \begin{cases} x, & \text{if } x \text{ is rational}, \\ -x, & \text{if } x \text{ is irrational}. \end{cases}$$

On can see that as $x$ converges to 2 then the corresponding value of $y$ converges to 2 if $x$ is rational and to $-2$ if $x$ is irrational. Thus $\lim_{x \to 2} y = \pm 2$. *Where is the mistake?*

### 3.3.8 Problem 24

Maybe the case that for a function $f$ we have $\lim_{x \to c^+} f(x) = l^{\pm}$, where $l^{\pm}$ denotes the convergence to $l$ both from above and below?

### 3.3.9 Problem 25

Maybe the case that for a function $f$ we have $\lim_{x \to c^+} f(x) = \pm \infty$, where $\pm \infty$ denotes the divergence both to $+\infty$ and to $-\infty$?

### 3.3.10 Problem 26

Is there a contradiction between the previous cases $\lim_{x \to c^+} f(x) = l^{\pm}$, $\lim_{x \to c^+} f(x) = \pm \infty$ discussed in Problems 24, 25, and the theorem on uniqueness of limits? (See Problem 23 for the statement.)

### 3.3.11   Problem 27

If the limit $\lim_{x\to\square}(f(x) + g(x))$ of a sum of functions exists, do the limits $\lim_{x\to\square} f(x)$, $\lim_{x\to\square} g(x)$ of the two functions exist?

### 3.3.12   Problem 28

Is it true that if $\lim_{x\to c} f(x) = l_1$ and $\lim_{x\to c} f(x) = l_2$ then $\lim_{x\to c}(f(x) + g(x)) = l_1 + l_2$?

### 3.3.13   Problem 29

If the limit $\lim_{x\to\square} f(x)g(x)$ of a product of functions exists, do the limits $\lim_{x\to\square} f(x)$, $\lim_{x\to\square} g(x)$ of the two functions exist?

### 3.3.14   Problem 30

If the limit $\lim_{x\to\square} \frac{f(x)}{g(x)}$ of the ratio of two functions exists, do the limits $\lim_{x\to\square} f(x)$, $\lim_{x\to\square} g(x)$ of the two functions exist?

### 3.3.15   Problem 31

If the limit $\lim_{x\to\square} f^2(x)$ of the square of a function exists, does the limit $\lim_{x\to\square} f(x)$ of the function exist?

### 3.3.16   Problem 32

If $\lim_{x\to\square} g(x) = 0$ then $\lim_{x\to\square} \frac{1}{g(x)} = \infty$?

### 3.3.17   Problem 33

If $\lim_{x\to\square} g(x) = \infty$ then $\lim_{x\to\square} \frac{1}{g(x)} = 0$?

### 3.3.18 Problem 34

If $\begin{cases} \lim_{x \to \square} g(x) = 0 \\ \lim_{x \to \square} f(x) = \infty \end{cases}$ then $\lim_{x \to \square} (1 + g(x))^{f(x)} = e$?

### 3.3.19 Problem 35

Is it true or false that if

$$\begin{cases} \lim_{x \to \square} g(x) = \infty, \\ \lim_{x \to \square} f(x) = \infty, \\ \text{there exists } I_\square \text{ such that } f(x) \le h(x) \le g(x) \text{ for all } x \in I_\square, \end{cases}$$

then $\lim_{x \to \square} h(x) = \infty$?

### 3.3.20 Problem 36

Is it true or false that if

$$\begin{cases} \lim_{x \to \square} f(x) = +\infty, \\ f(x) \le g(x), \end{cases}$$

then $\lim_{x \to \square} g(x) = +\infty$?

### 3.3.21 Problem 37

In the proof of the continuity of the sin function in Sect. 3.2.5, we have first proved that $\lim_{x \to 0^+} \cos x = 1$ by estimating the function $y = \cos x$ by means of the two functions $y = 1$ and $y = 1 - \sin x$, algebraic the first and transcendental the second (see Chap. 2 for definitions). Is it possible to obtain the same result by replacing the minorant function $y = 1 - \sin x$ with an algebraic function? (*Hint: use other trigonometric inequalities.*)

### 3.3.22 Problem 38

Is it true that a function defined in the whole of $\mathbb{R}$ can be continuous only at one point?

### 3.3.23   Problem 39

Give an example of a function $y = f(x)$ which is discontinuous at any point of $\mathbb{R}$ and such that its square function $y = f^2(x)$ is continuous in the whole of $\mathbb{R}$.

### 3.3.24   Problem 40

Define two functions $y = f(x)$ and $y = g(x)$ with $0 \in D(f), D(g)$, bounded above and below, with a discontinuity of the second kind at $x = 0$ and such that the function $y = f(x) + g(x)$ has a discontinuity of the third kind at $x = 0$.

### 3.3.25   Problem 41

Give the definition of a function defined on a countably infinite subset of $[0, 1]$, which has a discontinuity of the third kind at any point of its domain.

### 3.3.26   Problem 41*

Give the definition of a function defined on a countably infinite subset of $[0, 1]$, which has a discontinuity of the first kind at any point of its domain.

### 3.3.27   Problem 42

Give an example of a function defined on a set of the type $[a, b] \cap \mathbb{Q}$ such that $\lim_{x \to c} f(x) = +\infty$ for all $c \in [a, b]$.

### 3.3.28   Problem 43

It is well-known that:

i) There exist functions which become continuous after being re-defined at their points of discontinuity. For example, see Fig. 3.12.
ii) There exist functions which cannot be re-defined as above. For example, see Fig. 3.13.

**Fig. 3.12** Problem 43

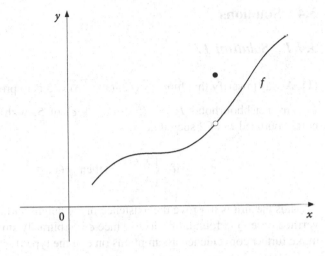

**Fig. 3.13** Problem 43

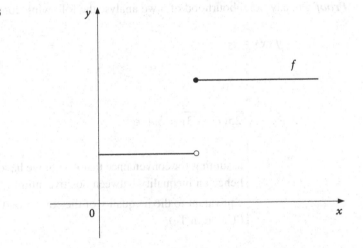

Give an example of a function which can be re-defined at its points of discontinuity in such a way that it becomes discontinuous *also* at other points.

## 3.3.29   *Problem 44*

Give an example of a function which can be re-defined at its points of discontinuity in such a way that it becomes continuous at those points and discontinuous at others.

## 3.4   Solutions

### 3.4.1   Solution 17

**(1)** A way to verify that $\lim_{x \to 2} \sqrt{25(x^2 - 3)} = 5$ is to prove that:

for any neighbourhood $I_5 = (5 - \epsilon, 5 + \epsilon)$ of 5, with $\epsilon > 0$, there exists a neighbourhood $I_2$ of 2 such that

$$\text{if } \begin{cases} x \in I_2, \\ x \neq 2 \end{cases} \text{ then } f(x) \in I_5 .$$

Thus the aim is to prove the existence of the neighbourhood $I_2$: the limit will be verified once $I_2$ is found. (To do so, since $\epsilon$ is arbitrarily small, it could be useful to make further convenience assumptions on $\epsilon$ of the type $0 < \epsilon < k$.)

**Proof** For any neighbourhood of 5 we analyse the following condition:

$$f(x) \in I_5$$

$$\updownarrow$$

$$5 - \epsilon < \sqrt{25(x^2 - 3)} < 5 + \epsilon$$

> assuming for convenience that $\epsilon < 5$, we have $5 - \epsilon > 0$, hence an inequality between positive numbers which is equivalent to the inequality of the corresponding squares (Theorem $T_8$)

$$(5 - \epsilon)^2 < 25(x^2 - 3) < (5 + \epsilon)^2$$

> multiplying by $\frac{1}{25}$ all terms in this inequality, we obtain an equivalent inequality (Theorem $T_3$)

$$\frac{(5 - \epsilon)^2}{25} < x^2 - 3 < \frac{(5 + \epsilon)^2}{25}$$

summing the number 3 to all terms in this inequality, we obtain an equivalent inequality (Theorem $T_1$)

$$\frac{(5-\epsilon)^2}{25} + 3 < x^2 < \frac{(5+\epsilon)^2}{25} + 3$$

it is not restrictive to consider only positive bases, hence it is possible to use Theorem $T_8$ which establishes that an inequality between positive numbers is equivalent to the inequality between the corresponding squares

$$\sqrt{\frac{(5-\epsilon)^2}{25} + 3} < x < \sqrt{\frac{(5+\epsilon)^2}{25} + 3}$$

$$x \in I = \left( \sqrt{4 - \frac{2}{5}\epsilon + \frac{\epsilon^2}{25}}, \sqrt{4 + \frac{2}{5}\epsilon + \frac{\epsilon^2}{25}} \right)$$

*Conclusion*

If $\begin{cases} x \in I = \left( \sqrt{4 - \frac{2}{5}\epsilon + \frac{\epsilon^2}{25}}, \sqrt{4 + \frac{2}{5}\epsilon + \frac{\epsilon^2}{25}} \right) \\ 0 < \epsilon < 5 \end{cases}$ then $f(x) \in I_5 = (5 - \epsilon, 5 + \epsilon)$.

It remains to prove that $I$ is a neighbourhood of 2, that is

$$\sqrt{4 - \frac{2}{5}\epsilon + \frac{\epsilon^2}{25}} < 2 < \sqrt{4 + \frac{2}{5}\epsilon + \frac{\epsilon^2}{25}}.$$

This is an easy exercise that can be performed by using the same calculations above.

(2) A way to verify that $\lim\limits_{x \to 1} \sqrt{\log_{\frac{1}{2}} \frac{1}{8x+8}} = 2$ is to prove that:

for any neighbourhood $I_2 = (2 - \epsilon, 2 + \epsilon)$ of 2, with $\epsilon > 0$, there exists a neighbourhood $I_1$ of 1 such that

$$\text{if } \begin{cases} x \in I_1, \\ x \neq 1 \end{cases} \text{ then } f(x) \in I_2.$$

Thus the aim is to prove the existence of the neighbourhood $I_1$: the limit will be verified once $I_1$ is found.

**_Proof_** For any neighbourhood of 2 we analyse the following condition:

$$f(x) \in I_2$$

$\updownarrow$

$$2 - \epsilon < \sqrt{\log_{\frac{1}{2}} \frac{1}{8x+8}} < 2 + \epsilon$$

$\uparrow$

assuming for convenience that $\epsilon < 2$, we have $2 - \epsilon > 0$, hence an inequality between positive numbers which is equivalent to the inequality of the corresponding squares (Theorem $T_8$)

$\downarrow$

$$(2-\epsilon)^2 < \log_{\frac{1}{2}} \frac{1}{8x+8} < (2+\epsilon)^2$$

$\uparrow$

the order of real numbers is equivalent to the opposite order of the corresponding powers with base $\frac{1}{2}$ (Theorem $T_{11}$)

$\downarrow$

$$\left(\frac{1}{2}\right)^{(2-\epsilon)^2} > \left(\frac{1}{2}\right)^{\log_{\frac{1}{2}} \frac{1}{8x+8}} > \left(\frac{1}{2}\right)^{(2+\epsilon)^2}$$

$\uparrow$

by definition of logarithm $a^{\log_a b} = b$

$\downarrow$

$$\frac{1}{2^{(2-\epsilon)^2}} > \frac{1}{8x+8} > \frac{1}{2^{(2+\epsilon)^2}}$$

$\uparrow$

an inequality between positive numbers is equivalent to the inequality of the corresponding inverses (Theorem $T_{14}$)

$\downarrow$

$$2^{(2-\epsilon)^2} < 8x + 8 < 2^{(2+\epsilon)^2}$$

$$\Big\uparrow \text{Theorems } T_1, T_3$$

$$\frac{2^{(2-\epsilon)^2}}{8} - 1 < x < \frac{2^{(2+\epsilon)^2}}{8} - 1$$

$$\uparrow$$

$$x \in I = \left( \frac{2^{(2-\epsilon)^2}}{8} - 1, \frac{2^{(2+\epsilon)^2}}{8} - 1 \right)$$

*Conclusion*

If $\begin{cases} x \in I = \left( \frac{2^{(2-\epsilon)^2}}{8} - 1, \frac{2^{(2+\epsilon)^2}}{8} - 1 \right) \\ 0 < \epsilon < 2 \end{cases}$ then $f(x) \in I_2 = (2 - \epsilon, 2 + \epsilon)$.

It remains to prove that $I$ is a neighbourhood of 2, that is

$$\frac{2^{(2-\epsilon)^2}}{8} - 1 < 2 < \frac{2^{(2+\epsilon)^2}}{8} - 1.$$

**(3)** A way to verify that $\lim\limits_{x \to 5} 9^{\sqrt{\frac{x-5}{x^2-6x+5}}} = 3$ is to prove that:

for any neighbourhood $I_3 = (\alpha, \beta)$ of 3, with $\alpha < 3 < \beta$, there exists a neighbourhood $I_5$ of 5 such that

$$\text{if } \begin{cases} x \in I_5, \\ x \neq 5 \end{cases} \text{ then } f(x) \in I_3.$$

Also in this case, it is useful to note that $\alpha$ and $\beta$ can be chosen arbitrarily close to 3. Thus, it will be possible to use convenience conditions of the type $h < \alpha < \beta < k$.

**Proof** For any neighbourhood of 3 we analyse the following condition:

$$f(x) \in I_3$$

$$\updownarrow$$

$$\alpha < 9\sqrt{\frac{x-5}{x^2-6x+5}} < \beta$$

$$\Bigg\uparrow$$
assuming for convenience that $\alpha > 0$, we have an inequality between positive numbers which is equivalent to the inequality of the corresponding logarithms with base 9 (Theorem $T_{12}$)

$$\log_9 \alpha < \log_9 \left( 9\sqrt{\frac{x-5}{x^2-6x+5}} \right) < \log_9 \beta$$

$$\updownarrow$$

$$\log_9 \alpha < \sqrt{\frac{x-5}{x^2-6x+5}} < \log_9 \beta$$

$$\Bigg\uparrow$$
assuming for convenience that $\alpha > 1$ we have $\log_9 \alpha > 0$, hence we have an inequality between positive numbers which is equivalent to the inequality of the corresponding squares (Theorem $T_8$)

$$\log_9^2 \alpha < \frac{x-5}{x^2-6x+5} < \log_9^2 \beta$$

$$\Big\uparrow$$ factoring the denominator

$$\log_9^2 \alpha < \frac{x-5}{(x-1)(x-5)} < \log_9^2 \beta$$

by definition of limit, we can assume that $x \neq 5$: under this assumption the previous inequality is equivalent to the one abtained by scaling out the factor $x - 5$

$$\log_9^2 \alpha < \frac{1}{x-1} < \log_9^2 \beta$$

the order of positive numbers is equivalent to the inverse order of their reciprocals (Theorem $T_{14}$)

$$\frac{1}{\log_9^2 \beta} < x - 1 < \frac{1}{\log_9^2 \alpha}$$

$$\frac{1}{\log_9^2 \beta} + 1 < x < \frac{1}{\log_9^2 \alpha} + 1$$

$$x \in I = \left( \frac{1}{\log_9^2 \beta} + 1, \frac{1}{\log_9^2 \alpha} + 1 \right)$$

*Conclusion*

If $\begin{cases} x \in I = \left( \frac{1}{\log_9^2 \beta} + 1, \frac{1}{\log_9^2 \alpha} + 1 \right) \\ x \neq 5 \\ \alpha > 1 \end{cases}$ then $f(x) \in I_3 = (\alpha, \beta)$.

It remains to prove that $I$ is a neighbourhood of 5, that is

$$\frac{1}{\log_9^2 \beta} + 1 < 5 < \frac{1}{\log_9^2 \alpha} + 1.$$

**(4)** We have to prove that:

for any neighbourhood $I_{+\infty} = (M, +\infty)$ of $+\infty$, there exists a right neighbourhood $I_{1+}$ of 1 such that if $x \in I_{1+}$ then $f(x) \in I_{+\infty}$.

We note that $M$ can be chosen arbitrarily large. Thus, it will be possible to use convenience conditions of the type $M > k$.

***Proof*** For any neighbourhood of $+\infty$ we analyse the following condition:

$$f(x) \in I_{+\infty}$$

$\updownarrow$

$$2^{|\log_x 3|} > M$$

assuming for convenience that $M > 0$, we have an inequality between positive numbers which is equivalent to the inequality of the corresponding logarithms with base 2 (Theorem $T_{12}$)

$$|\log_x 3| > \log_2 M$$

since we are considering $x \to 1^+$, it is possible to assume that $x > 1$, which implies that $\log_x 3 > 0$ hence $|\log_x 3| = \log_x 3$

$$\log_x 3 > \log_2 M$$

recalling that $\log_x 3 = (\log_3 x)^{-1}$

$$\frac{1}{\log_3 x} > \log_2 M$$

assuming for convenience that $M > 1$, we have $\log_2 M > 0$, hence we have an inequality between positive numbers which is equivalent to the opposite inequality of the corresponding reciprocals (Theorem $T_{14}$)

$$0 < \log_3 x < (\log_2 M)^{-1}$$

Theorem $T_{10}$

$$1 < x < 3^{(\log_2 M)^{-1}}$$

$$x \in I_{1+} = \left(1, 3^{(\log_2 M)^{-1}}\right)$$

*Conclusion*

If $\begin{cases} x \in I_{1+} = \left(1, 3^{(\log_2 M)^{-1}}\right) \\ M > 1 \end{cases}$ then $f(x) \in I_{+\infty} = (M, +\infty)$.

**(5)** We have to prove that:

for any neighbourhood $I_{-\infty} = (-\infty, M)$ of $-\infty$, there exists a right neighbourhood $I_{+\infty}$ of $+\infty$ such that if $x \in I_{+\infty}$ then $f(x) \in I_{-\infty}$.

We note that the goal does not consist in solving the inequality $f(x) < M$ but only to prove that the set of its solutions contains a neighbourhood of $+\infty$. Thus, it will be possible to solve that inequality in intervals of the type $(k, +\infty)$, which means that we can use convenience assumptions of the type $x > k$.

**Proof** For any neighbourhood of $-\infty$ we analyse the following condition:

$$f(x) \in I_{-\infty}$$

$$\sqrt[3]{\log_2 \sin \frac{1}{x}} < M$$

Theorem T$_7$

$$\log_2 \sin \frac{1}{x} < M^3$$

the inequality between two numbers is equivalent to the
inequality between the corresponding powers with base 2
(Theorem T$_{10}$)

$$\sin \frac{1}{x} < 2^{M^3}$$

recalling the trigonometric inequality $\sin \alpha < \alpha$ if $\alpha > 0$
we have that $\sin \frac{1}{x} < \frac{1}{x}$ if $\frac{1}{x} > 0$, that is $x > 0$. Thus, if
$0 < \frac{1}{x} < 2^{M^3}$ then $\sin \frac{1}{x} < 2^{M^3}$. See Fig. 3.14.

$$0 < \frac{1}{x} < 2^{M^3}.$$

Theorem T$_{14}$

$$x > 2^{-M^3}$$

$$x \in I_{+\infty} = \left(2^{-M^3}, +\infty\right)$$

*Conclusion*

If $x \in I_{+\infty} = \left(2^{-M^3}, +\infty\right)$ then $f(x) \in I_{-\infty} = (-\infty, M)$.

Question: Is $I_{+\infty} = \left(2^{-M^3}, +\infty\right)$ the unique neighbourhood of $+\infty$ such that
$f(x) \in I_{-\infty}$ for all $x \in I_{+\infty}$? Answer: No! Any neighbourhood contained in $I_{+\infty}$
of the type $I = (\alpha, +\infty)$ with $\alpha \geq 2^{-M^3}$ satisfies that condition.

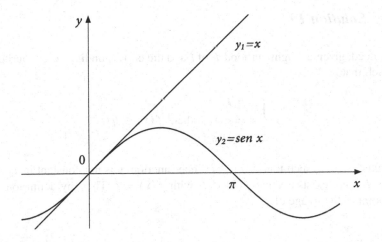

**Fig. 3.14** Solution 17. If $x \geq 0$ then $\sin x \leq x$

## 3.4.2 Solution 18

Not always.

*Example* $y = 1$

See Fig. 3.15.

Indeed, Im $(f) = \{1\}$ is a finite set (it contains only one point!), hence it has no limit points.

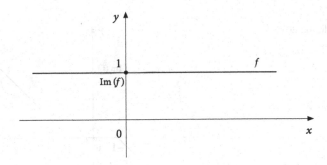

**Fig. 3.15** Solution 18

### 3.4.3   Solution 19

Yes. Indeed, given a neighbourhood $I_l$ of $l$ and the corresponding neighbourhood $I_c$ of $c$ such that

$$\text{if } \begin{cases} x \in I_c \\ x \neq c \\ x \in D(f) \end{cases} \text{ then } f(x) \in I_l \ ,$$

there exists $\bar{x} \in I_c$ such that $f(\bar{x}) \neq l$ since function $f$ is not constant in $I_c$. Thus, for any $I_l$ there exists a value $f(\bar{x}) \in I_l$ with $f(\bar{x}) \neq l$. Thus, by definition, $l$ is a limit point of the image of $f$.

### 3.4.4   Solution 20

Not necessarily.

*Example*

$$f(x) = \begin{cases} x - 1, & \text{if } x \geq 1, \\ 0, & \text{if } -1 < x < 1, \\ x + 1, & \text{if } x \leq -1. \end{cases}$$

See Fig. 3.16.

**Fig. 3.16**  Solution 20

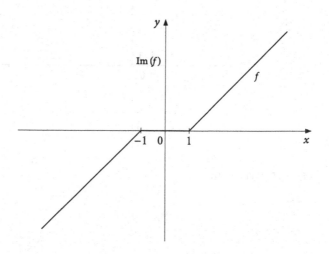

Indeed $\lim_{x \to 0} f(x) = 0$, the image of $f$ is $\mathbb{R}$, hence 0 is a limit point of the image of $f$. However, $f$ is constant on $[-1, 1]$.

### 3.4.5 Solution 21

Not necessarily.

*Example*

$$f(x) = \begin{cases} x, & \text{if } x \text{ is rational,} \\ -x, & \text{if } x \text{ is not rational.} \end{cases}$$

See Fig. 3.17.

Indeed, the image of $f$ is $\mathbb{R}$, hence any $l \in \mathbb{R}$ is a limit point of the image of $f$. However, if $l \neq 0$, there does not exist $c \in \mathbb{R}$ such that $\lim_{x \to c} f(x) = l$.

**Fig. 3.17** Solution 21

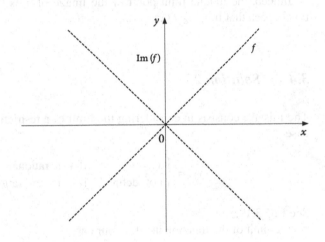

### 3.4.6 Solution 22

Not necessarily.

*Example*

$$f(x) = \begin{cases} 1, & \text{if } x \notin \{\frac{1}{n} : n \in \mathbb{N}_0\}, \\ x, & \text{if } x \in \{\frac{1}{n} : n \in \mathbb{N}_0\}. \end{cases}$$

See Fig. 3.18.

**Fig. 3.18** Solution 22

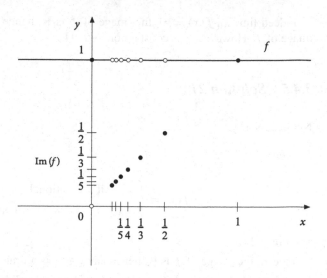

Indeed, the unique limit point of the image of $f$ is 0 but there does not exist $c \in \mathbb{R}$ such that $\lim_{x \to c} f(x) = 0$.

### 3.4.7  Solution 23

The mistake consists in considering the limit of a restriction of the function, in our case

$$y_1 = \begin{cases} x, & \text{if } x \text{ is rational} , \\ \text{not defined}, & \text{if } x \text{ is not rational} , \end{cases}$$

See Fig. 3.19.
As the limit of the function itself, in our case

$$y = \begin{cases} x, & \text{if } x \text{ is rational} , \\ -x, & \text{if } x \text{ is not rational} . \end{cases}$$

It is true that $\lim_{x \to 2} y_1 = 2$ but it is not true that $\lim_{x \to 2} y = 2$.

Indeed, given the neighbourhood $I_2 = (1, 3)$ of 2 in the $y$-axis, even if we consider a neighbourhood of 2 in the $x$-axis as small as possible, we shall always find points of the graph of $f$ which belong to the fourth quadrant, hence outside the strip $1 < y < 3$ (see Fig. 3.20). Thus, the limits of the two restrictions to rational and irrational numbers

$$\lim_{\substack{x \to 2, \\ x \in \mathbb{Q}}} y = 2, \quad \lim_{\substack{x \to 2, \\ x \notin \mathbb{Q}}} y = -2$$

exist, but the limit $\lim_{x \to 2} y = 2$ does not exist. Thus, the theorem is safe!

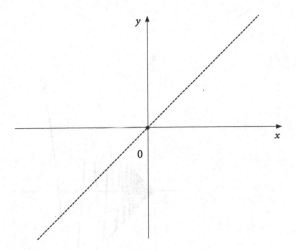

**Fig. 3.19** Solution 23

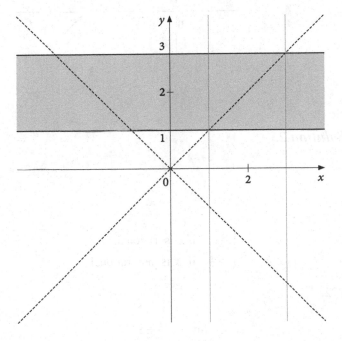

**Fig. 3.20** Solution 23

### 3.4.8   Solution 24

Yes.

*Example*  $\lim_{x \to 0^+} \left(2 + x \sin \frac{1}{x} = 2^{\pm}\right)$
See Fig. 3.21.

**Fig. 3.21**  Solution 24

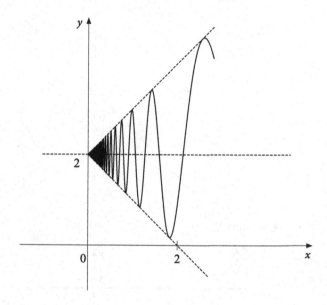

### 3.4.9   Solution 25

Yes.

*Example*

$$y = \begin{cases} \frac{1}{x}, & \text{if } x \text{ is rational,} \\ -\frac{1}{x}, & \text{if } x \text{ is not rational .} \end{cases}$$

See Fig. 3.22.

$$\lim_{x \to 0^+} y = \pm\infty$$

In particular:

$$\lim_{\substack{x \to 0^+, \\ x \in \mathbb{Q}}} y = +\infty, \quad \lim_{\substack{x \to 0^+, \\ x \notin \mathbb{Q}}} y = -\infty, \quad \lim_{\substack{x \to 0^+, \\ x \in \mathbb{R}}} y = \pm\infty.$$

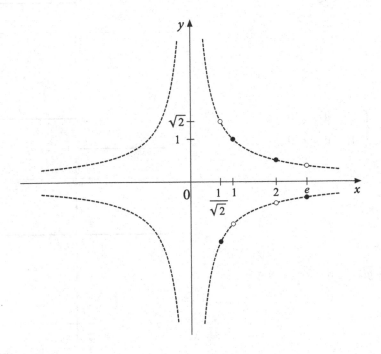

**Fig. 3.22** Solution 25

### 3.4.10  Solution 26

No, because that theorem establishes the uniqueness of the limit and not the way the limit is reached (by above or below). In this sense, $+\infty$ and $-\infty$ can be considered as the two possible ways of diverging to $\infty$.

### 3.4.11  Solution 27

Not necessarily.

*Example* See Fig. 3.23.

$$f(x) = \begin{cases} 1, & \text{if } x \geq 0, \\ 2, & \text{if } x < 0. \end{cases}$$

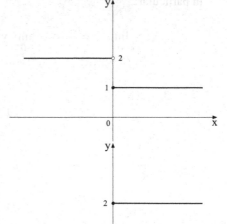

$$g(x) = \begin{cases} 2, & \text{if } x \geq 0, \\ 1, & \text{if } x < 0. \end{cases}$$

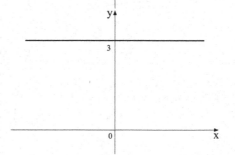

$$f(x) + g(x) = 3, \quad \text{for all } x \in \mathbb{R}$$

**Fig. 3.23** Solution 27

In this case:

$\lim_{x \to 0} f(x)$ does not exist,

$\lim_{x \to 0} g(x)$ does not exist,

but

$\lim_{x \to 0} f(x) + g(x) = \lim_{x \to 0} 3 = 3.$

### 3.4.12  Solution 28

Not necessarily. The limit of the sum may not exist since the sum of the two functions may not be defined.

*Example* $f(x) = 1$ if $x$ is rational (see Fig. 3.24).

$g(x) = 1$ if $x$ is not rational (see Fig. 3.25).

The domain of $f(x) + g(x)$ is empty, hence it does not make sense to talk about the limit of the sum.

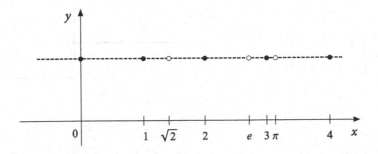

**Fig. 3.24** Solution 28

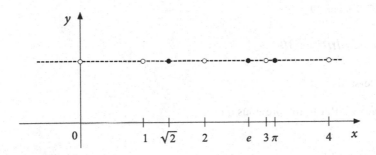

**Fig. 3.25** Solution 28

### 3.4.13 Solution 29

Not necessarily.

*Example* See Fig. 3.26.

Indeed, $\lim_{x \to 0} f(x)g(x) = 0$ but the two limits $\lim_{x \to 0} f(x)$ and $\lim_{x \to 0} fg(x)$ do not exist.

$$f(x) = \begin{cases} 0, & \text{if } x \le 0, \\ 1, & \text{if } x > 0. \end{cases}$$

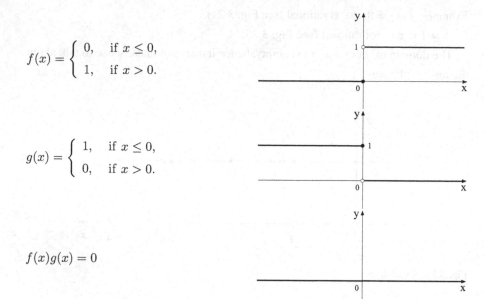

$$g(x) = \begin{cases} 1, & \text{if } x \le 0, \\ 0, & \text{if } x > 0. \end{cases}$$

$$f(x)g(x) = 0$$

**Fig. 3.26** Solution 29

### 3.4.14  Solution 30

Not necessarily.

*Example* Similar to the previous one.

### 3.4.15  Solution 31

Not necessarily.

*Example*

$$f(x) = \begin{cases} 1, & \text{if } 2n \le x \le 2n + 1, \\ -1, & \text{if } 2n + 1 < x < 2n + 2, \end{cases}$$

where $n$ assumes all values in $\mathbb{N}$. See Fig. 3.27.

We have that $\lim_{x \to +\infty} f(x)$ does not exist, but since $f^2(x) = 1$ for all $x \in \mathbb{R}$ with $x \ge 0$ we have that

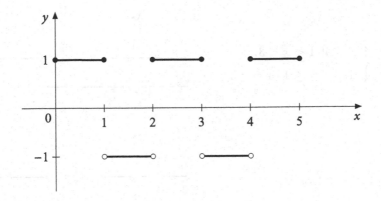

**Fig. 3.27** Solution 31

$$\lim_{x \to +\infty} f^2(x) = \lim_{x \to +\infty} 1 = 1.$$

In the same way, we have that for any $n \in \mathbb{N}$, $n \neq 0$, $\lim_{x \to n} f(x)$ does not exist, whilst $\lim_{x \to n} f^2(x) = \lim_{x \to n} 1 = 1$.

### 3.4.16 Solution 32

Not always. It may happen that $y = \frac{1}{g(x)}$ is not defined at all or that $\square$ is not a limit point of the domain of $y = \frac{1}{g(x)}$.

*Example 1*
If $g(x) = 0$ for all $x \in \mathbb{R}$ then $y = \frac{1}{g(x)}$ is not defined.

*Example 2*
See Fig. 3.28.
In this case we have that $\lim_{x \to 2} g(x) = 0$ whilst $\lim_{x \to 2} \frac{1}{g(x)}$ is not well-defined, because the point $x = 2$ is not a limit point of the domain of $\frac{1}{g(x)}$.

### 3.4.17 Solution 33

Yes.

$$g(x) = \begin{cases} 0, & \text{if } 1 < x < 3, \\ 1, & \text{if } x \le 1 \text{ or } x \ge 3, \end{cases}$$

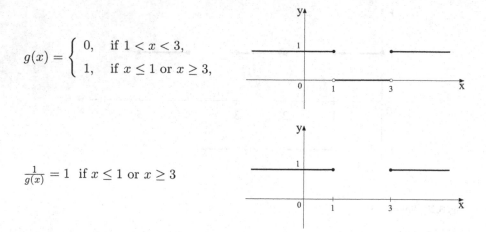

$$\frac{1}{g(x)} = 1 \ \text{ if } x \le 1 \text{ or } x \ge 3$$

**Fig. 3.28** Solution 32

### 3.4.18   Solution 34

Yes but only if we also assume in addition that $f(x) = \frac{1}{g(x)}$ for all $x \in I_\square$ with $x \ne \square$, for a suitable neighborhood $I_\square$ of $\square$.

### 3.4.19   Solution 35

False.

*Example* We set $\mathcal{A} = \{\frac{1}{n} : n \in \mathbb{N}_0\}$ and we define the functions on the positive real semi-axis (see Figs. 3.29 and 3.30):

$$f(x) = \begin{cases} \frac{1}{x}, & \text{if } x \text{ is rational,} \\ -\frac{1}{x}, & \text{if } x \text{ is not rational,} \end{cases} \qquad g(x) = \begin{cases} \frac{1}{x}, & \text{if } x \notin \mathcal{A}, \\ -\frac{1}{x}, & \text{if } x \in \mathcal{A}. \end{cases}$$

$$h(x) = \begin{cases} \frac{1}{x}, & \text{if } x \text{ is not rational,} \\ 0, & \text{if } x \text{ is rational and } x \notin \mathcal{A}, \\ -\frac{1}{x}, & \text{if } x \in \mathcal{A}. \end{cases}$$

It is evident that

$$\lim_{x \to 0^+} f(x) = \lim_{x \to 0^+} g(x) = \infty$$

and on the other hand $f(x) \le h(x) \le g(x)$. Indeed:

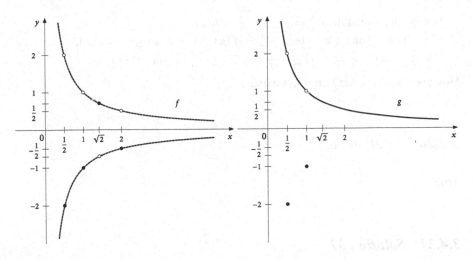

**Fig. 3.29** Solution 35

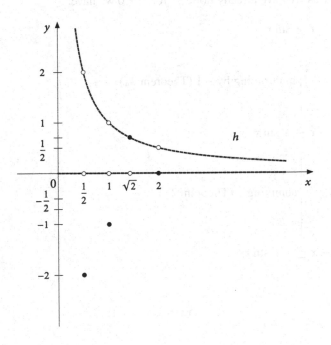

**Fig. 3.30** Solution 35

- If $x$ is not rational then $\frac{1}{x} = f(x) \le \frac{1}{x} = h(x) \le \frac{1}{x} = g(x)$.
- If $x$ is rational and $x \notin \mathcal{A}$ then $-\frac{1}{x} = f(x) \le 0 = h(x) \le \frac{1}{x} = g(x)$.
- If $x \in \mathcal{A}$ then $-\frac{1}{x} = f(x) \le -\frac{1}{x} = h(x) \le -\frac{1}{x} = g(x)$.

However, $\lim_{x \to 0^+} h(x)$ does not exist.

### 3.4.20  Solution 36

True.

### 3.4.21  Solution 37

Yes. Indeed, as we have already noticed, for $x \ge 0$ we have:

$$x \ge \sin x$$

multiplying by $-1$ (Theorem $T_4$)

$$-x \le -\sin x$$

summing 1 (Theorem $T_1$)

$$1 - x \le 1 - \sin x$$

and

$$1 - \sin x \le \cos x \le 1.$$

Thus, by transitivity we get

$$1 - x \le \cos x \le 1.$$

Passing to the limit as $x \to 0^+$ we get $\lim_{x \to 0^+} \cos x = 1$:

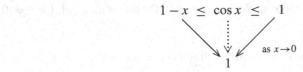

## 3.4.22 Solution 38

Yes.

*Example 1*

$$y = \begin{cases} x, & \text{if } x \text{ is rational}, \\ -x, & \text{if } x \text{ is not rational}. \end{cases}$$

See Fig. 3.31.

**Fig. 3.31** Solution 38
(Example 1)

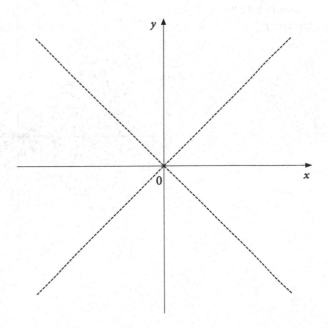

By applying the Squeeze Theorem with squeezing functions $y = x$ and $y = -x$, we have that

$$\lim_{x \to 0} f(x) = 0 = f(0),$$

hence the function is continuous at the point $x = 0$. On the other hand, for $c \neq 0$ we have that

$$\lim_{\substack{x \to c \\ x \in \mathbb{Q}}} f(x) = \lim_{x \to c} x = c \quad \text{and} \quad \lim_{\substack{x \to c \\ x \notin \mathbb{Q}}} f(x) = \lim_{x \to c} -x = -c,$$

hence $\lim_{x \to c} f(x)$ doesn't exist. Thus, the function is continuous only at the point $x = 0$.

*Example 2*
Consider the 'fog function':

$$y = \text{random}\,(x), \quad \text{with} \; -|x| \leq \text{random}\,(x) \leq |x|, \quad \text{for all } x \in \mathbb{R},$$

where random $(x)$ denotes a random number associated with $x$. See Fig. 3.32.

**Fig. 3.32** Solution 38 (Example 2: the 'fog function')

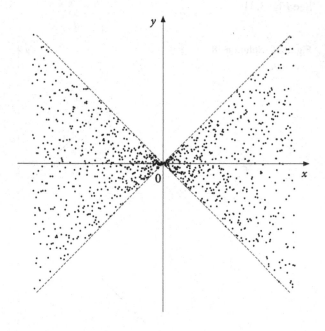

Note that $f(0) = 0$ because by definition we have $0 \leq$ random $(0) \leq 0$. Moreover, $\lim_{x \to 0}$ random $(x) = 0$ by the Squeeze Theorem applied with squeezing functions $y = |x|$ and $y = -|x|$:

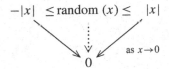

Thus the function is continuous at $x = 0$. However, although the graph in Fig. 3.32 has been randomly generated and would confirm our intuition that a function like that can be continuous only at $x = 0$, one cannot use this function to answer the question in a rigorous way. On the other hand, it suffices to modify it as follows:

$$y = \begin{cases} \text{random } (x) \in [x^2, 4x^2], & \text{if } x \text{ is rational}, \\ \text{random } (x) \in [-4x^2, -x^2], & \text{if } x \text{ is irrational}. \end{cases}$$

This function is certainly continuous only at $x = 0$.

### 3.4.23 Solution 39

*Example*

$$y = \begin{cases} 1, & \text{if } x \text{ is rational }, \\ -1, & \text{if } x \text{ is not rational}. \end{cases}$$

$$y = f^2(x) = 1, \quad \text{for all } x \in \mathbb{R}.$$

### 3.4.24 Solution 40

*Particular example*

$$f(x) = \begin{cases} \sin^2 \frac{1}{x}, & \text{if } x \neq 0, \\ 0, & \text{if } x = 0, \end{cases} \qquad g(x) = \begin{cases} \cos^2 \frac{1}{x}, & \text{if } x \neq 0, \\ 0, & \text{if } x = 0, \end{cases}$$

We have that $f(x) + g(x) = 1$ for all $x \neq 0$, $\lim_{x \to 0} f(x)$ and $\lim_{x \to 0} g(x)$ do not exist but $\lim_{x \to 0}(f(x) + g(x)) = 1 \neq f(0) + g(0) = 0$.

*General example*
Let $f(x)$ be a function with the required properties. Then, in order to answer to the question, it suffices to set $g(x) = 1 - f(x)$ and note that $\lim_{x \to 0}(f(x) + g(x)) = 1$.

### 3.4.25  Solution 41

*Example* We construct the function step by step with the following iterative procedure:

STEP 1.  We consider a 'bounding' square with sides of length $l$ and parallel to the coordinate axis, the lower basis of which has $y$-coordinate equal to $b$ and the right vertexes of which have $x$-coordinate equal to $v$ (Fig. 3.33).

**Fig. 3.33** Solution 41, step 1

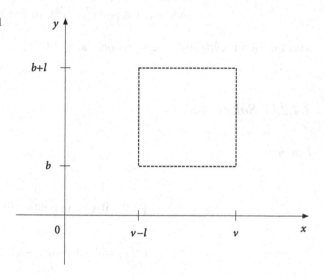

STEP 2.  We define the subset $D_v = \left\{ v - \frac{l}{2}, v - \frac{l}{4}, v - \frac{l}{8}, \ldots, v - \frac{l}{2^n}, \ldots \right\}$ of the domain of the function, which has an accumulation point at $x = v$.

STEP 3.  We set $f(x) = b + l$ for all $x \in D_v$ (Fig. 3.34).

STEP 4.  We construct 'bounding sub-squares' with sides (parallel to the coordinate axis) of lengths $\frac{l}{2}, \frac{l}{4}, \frac{l}{8}, \ldots, \frac{l}{2^n}, \ldots$ the lower basis of which have $y$-coordinates equal to $b, b + \frac{l}{2}, b + \frac{l}{2} + \frac{l}{4}, b + \frac{l}{2} + \frac{l}{4} + \frac{l}{8}, \ldots, b + \frac{l}{2} + \frac{l}{4} + \frac{l}{8} + \cdots + \frac{l}{2^n}, \ldots$ respectively, and the right vertexes of which have $x$-coordinates equal to $v - \frac{l}{2}, v - \frac{l}{4}, v - \frac{l}{8}, \ldots, v - \frac{l}{2^n}, \ldots$ respectively (Fig. 3.35).

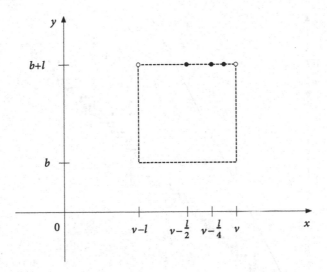

**Fig. 3.34**  Solution 41, step 3

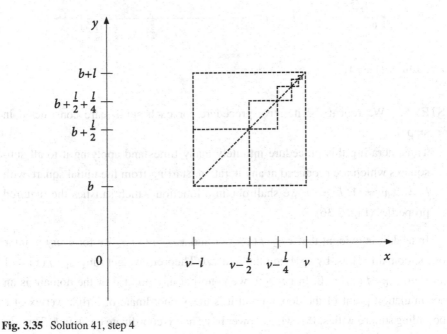

**Fig. 3.35**  Solution 41, step 4

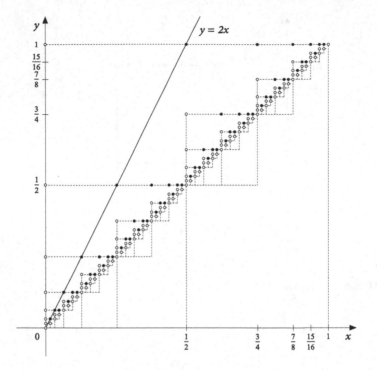

**Fig. 3.36**  Solution 41

STEP 5.    We repeat the iterative procedure for each sub-square constructed in
step 3.

Thus, iterating this procedure infinitely many times and applying it to all sub-
squares which are produced at any iteration, starting from the initial square with
$l = 1$, $v = 1$, $b = 0$, we shall obtain a function which satisfies the required
properties (Fig. 3.36).

Indeed, it is evident that $x \leq f(x) \leq 1$ and $x \leq f(x) \leq 2x$ for every point $x$
of the domain. Thus, by applying the Squeeze Theorem we get $\lim_{x \to 1^-} f(x) = 1$
and $\lim_{x \to 0^+} f(x) = 0$. In general, we note that any point $v$ of the domain is an
accumulation point of the domain and it is the $x$-coordinate of a right vertex of a
bounding square with side $l$, whose lower basis has $y$-coordinate equal to $b$. Thus, in
order to compute $\lim_{x \to v^-} f(x)$ it suffices to consider the values of $x$ with $v - l <$
$x < v$: for these values we have $x \leq f(x) \leq b + l$. Thus by passing to the limit
it follows that $\lim_{x \to v^-} f(x) = v$ since $b + l = v$. On the other hand, in order to
to compute $\lim_{x \to v^+} f(x)$, it suffices to consider the values of $x$ with $x > v$: for

these values we have that $x \le f(x) \le 2x - v$. Thus, by passing to the limit we have $\lim_{x \to v+} f(x) = v$. In conclusion, we have that $\lim_{x \to v} f(x) = v$ but $v \ne f(v)$. For example, with $v = \frac{1}{2}, b = 0, l = \frac{1}{2}$, we have $\lim_{x \to \frac{1}{2}} f(x) = \frac{1}{2}$ but $f(\frac{1}{2}) = 1$.

### 3.4.26  Solution 41*

*Example* We construct the function step by step with the following iterative procedure:

STEP 1.  We consider a 'bounding' rectangle with sides of length $b$ and $h$ and parallel to the coordinate axis, the right vertexes of which have $x$-coordinate equal to $v$ and the upper vertexes of which have $y$-coordinate equal to $w$ (Fig. 3.37).

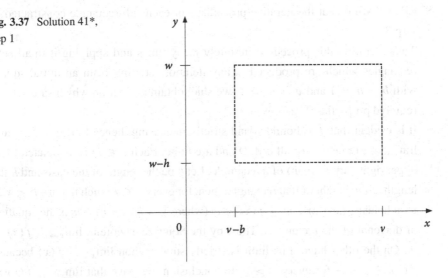

**Fig. 3.37**  Solution 41*, step 1

STEP 2.  We define the subset $D_v = \left\{ v - \frac{b}{2}, v - \frac{b}{4}, v - \frac{b}{8}, \ldots, v - \frac{b}{2^n}, \ldots \right\}$ of the domain of the function, which has an accumulation point at $x = v$.

STEP 3.  We set $f\left(v - \frac{b}{2^n}\right) = w - \frac{h}{2^n}$ for all $n = 1, 2, 3, \ldots$.

STEP 4.  We consider 'bounding sub-rectangles' with sides parallel to the $x$-axis of lengths $\frac{b}{2}, \frac{b}{4}, \frac{b}{8}, \ldots, \frac{b}{2^n}, \ldots$, sides parallel to the $y$-axis of lengths $\frac{h}{4}, \frac{h}{8}, \frac{h}{16}, \ldots, \frac{h}{2^{n+1}}, \ldots$, and the upper right vertexes of which have coordinates equal to $(v - \frac{b}{2}, w - \frac{h}{2}), (v - \frac{b}{4}, w - \frac{h}{4}), (v - \frac{b}{8}, w - \frac{h}{8}), \ldots, (v - \frac{b}{2^n}, w - \frac{h}{2^n}), \ldots$ respectively (Fig. 3.38).

**Fig. 3.38** Solution 41*, step 4.

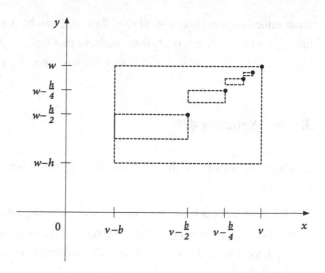

STEP 5.    We repeat the iterative procedure for each sub-rectangle constructed in step 3.

Thus, iterating this procedure infinitely many times and applying it to all sub-rectangles which are produced at any iteration, starting from an initial square with $b = h = 1$ and $v = w = 1$, we shall obtain a function which satisfies the required properties (Fig. 3.39).

It is evident that $f$ is bounded and strictly increasing, hence $\lim_{x \to c^-} f(x)$ and $\lim_{x \to c^+} f(x)$ exist for all $c \in D$ and are finite. Each $v \in D$ is associated to an upper right vertex $(v, w)$ of a rectangle. Let $b$ be the length of the basis and $h$ the length of the height of that rectangle: then, for every $x \in D$ such that $v - b < x < v$, it holds: $\frac{h}{b}(x - v) + w \le f(x) < w$ (where $y = \frac{h}{b}(x - v) + w$ is the equation of diagonal of the rectangle). Thus, by the Squeeze Theorem, $\lim_{x \to v^-} f(x) = w$. On the other hand, this limit is strictly smaller than $\lim_{x \to v^+} f(x)$ because $f(x) > w + \frac{h}{2}$ for every $x > v$. In conclusion, we have that $\lim_{x \to v^-} f(x) = w < \lim_{x \to v^+} f(x)$ for each $v \in D$.

### 3.4.27   Solution 42

*Example 1*
We recall that every rational number can be written as an irreducible fraction where the numerator $p$ and the denominator $q$ are relatively prime. Moreover, this fraction can be written in two irreducible forms:

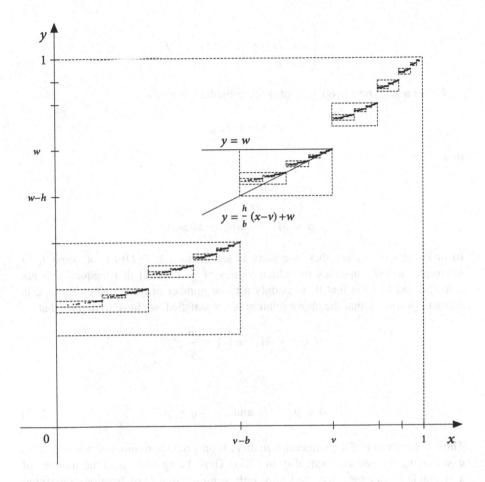

**Fig. 3.39** Solution 41*

$$\frac{p}{q} \quad \text{and} \quad \frac{-p}{-q}.$$

In our case, we consider rational numbers belonging to the interval $[1, 2]$ and written in the following form: $\frac{p}{q} \in [1, 2]$ with $p, q$ relatively prime and $p > 0$.

Then, we define a function on the set $[1, 2] \cap \mathbb{Q}$ as follows:

$$f\left(\frac{p}{q}\right) = p.$$

We now prove that $\lim_{x \to c} f(x) = +\infty$ for all $c \in [1, 2]$. To do so, we have to prove that for any neighbourhood $I_{+\infty}$ of $+\infty$ there exists a neighbourhood $I_c$ of $c$ such that

$$\text{if } \begin{cases} x \in I_c \\ x \in D(f) \quad \text{then} \quad f(x) \in I_{+\infty} \\ x \neq c \end{cases}$$

Given a neighbourhood $I_{+\infty}$ of $+\infty$, consider the relation

$$f(x) \in I_{+\infty}$$

that is

$$
\begin{aligned}
f(x) &> M \\
f\left(\frac{p}{q}\right) &> M \\
p &> M, \qquad \text{where } M > 0.
\end{aligned}
$$

In order to see whether this inequality is satisfied on $I_c \cap D(f)$ for some $I_c$, it is simpler to ask ourselves for which values of $\frac{p}{q} \in [1, 2]$ this inequality is not satisfied and find out that there is only a finite number of those values. In fact, in order to guarantee that the above relation is not satisfied, we have to require that

$$0 < p \leq M, \text{ and } 1 \leq \frac{p}{q} \leq 2$$

that is

$$0 < p \leq M, \text{ and } \frac{p}{2} \leq q \leq p. \tag{3.7}$$

Thus, for any value of the numerator $p$, there is only a finite number of denominators $q$ satisfying the second inequality in (3.7). Thus, being finite also the number of admissible numerators, we shall have only a finite number of fractions satisfying (3.7). Thus, since the condition $f(x) \in I_{+\infty}$ is violated at most by a finite number of rational numbers in $[1, 2]$, it will be possible to find a neighbourhood $I_c$ such that $f(x) \in I_{+\infty}$ for all rational numbers in that neighbourhood different from $c$, see Figs. 3.40 and 3.41. For example, if the rational numbers that violate the condition $f(x) \in I_{+\infty}$ are $x_1, x_2, x_3, x_4, x_5$, then the neighbourhood of $c$ can be defined by

$$I_c = (c - \delta, c + \delta)$$

where

$$\delta = \min\{|x_1 - c|, |x_2 - c|, |x_3 - c|, |x_4 - c|, |x_5 - c|\}.$$

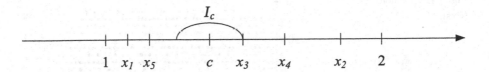

**Fig. 3.40**  Solution 42

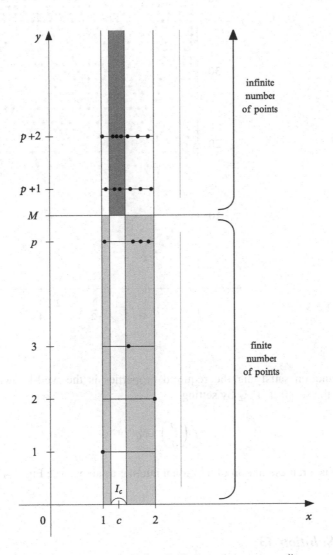

**Fig. 3.41**  Solution 42 (Example 1). "Scala Paradisii": the staircase to paradise

**Fig. 3.42** Solution 42
(Example 2)

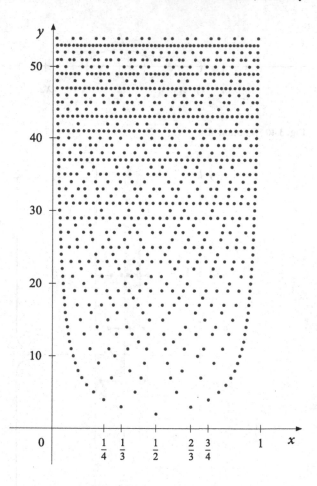

*Example 2*
Another function satisfying the required properties is the well-known function
defined on the set $(0, 1) \cap \mathbb{Q}$ by setting

$$f\left(\frac{p}{q}\right) = q$$

where $p/q$ is a representation of a rational number as above, see Fig. 3.42.

### 3.4.28   Solution 43

*Example* We define the following functions (see Figs. 3.43 and 3.44).

**Fig. 3.43** Solution 43

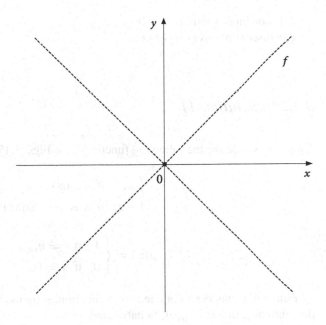

**Fig. 3.44** Solution 43

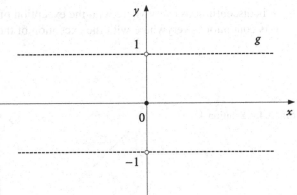

$$f(x) = \begin{cases} x, & \text{if } x \text{ is rational,} \\ -x, & \text{if } x \text{ is not rational.} \end{cases}$$

$$g(x) = \begin{cases} 1, & \text{if } x \text{ is rational with } x \neq 0, \\ 0, & \text{if } x = 0, \\ -1, & \text{if } x \text{ is not rational.} \end{cases}$$

Function $g$ has been obtained from function $f$ by modifying $f$ at its points of discontinuity, that is $\mathbb{R} \setminus \{0\}$. In particular:

– $f$ is continuous only at $x = 0$
– $g$ is discontinuous everywhere.

### 3.4.29  Solution 44

*Example* We define the following functions (see Figs. 3.45 and 3.46).

$$f(x) = \begin{cases} x, & \text{if } x \text{ is rational,} \\ -x, & \text{if } x \text{ is not rational.} \end{cases}$$

$$g(x) = \begin{cases} 1, & \text{if } x \neq 0, \\ 0, & \text{if } x = 0. \end{cases}$$

Function $g$ has been obtained from function $f$ by modifying $f$ at its points of discontinuity, that is $\mathbb{R} \setminus \{0\}$. In particular:

– $f$ is discontinuous everywhere with the exception of the point $x = 0$
– $g$ is continuous everywhere with the exception of the point $x = 0$.

**Fig. 3.45** Solution 44

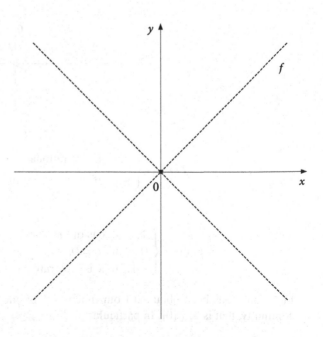

**Fig. 3.46** Solution 44

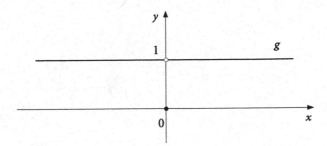

# Chapter 4
# Differentiation

## 4.1 Theoretical background

In this chapter, we consider the notion of derivative for functions of one real variable. Although its definition can be found in any textbook, we believe it is useful to state it here in the spirit of our general framework:

**Definition 4.1** Let $\mathcal{A}$ be a subset of $\mathbb{R}$ and $f$ a function from $\mathcal{A}$ to $\mathbb{R}$. Assume that $c \in \mathcal{A}$ is a limit point of $\mathcal{A}$. The derivative of $f$ at $c$ is denoted by $f'(c)$ and is defined by

$$f'(c) = \lim_{x \to c} \frac{f(x) - f(c)}{x - c}, \tag{4.1}$$

whenever that limit exists. Moreover, we say that $f$ is differentiable at $c$ if $f'(c)$ exists and is finite. Finally, the derivative (function) of $f$ is the function $f'$ defined on the set $D(f') = \{x \in \mathcal{A} : f \text{ is differentiable at } x\}$ and which takes any $x \in D(f')$ to the derivative $f'(x)$ of $f$ at $x$.

The derivative of a function $y = f(x)$ is also denoted by

$$y' \quad \text{or} \quad \frac{dy}{dx}.$$

Note that the limit in (4.1) could be equal to $+\infty$, $-\infty$ or $\infty$, in which case $f$ is not considered to be differentiable. For example, for the function $f$ from $\mathbb{R}$ to itself defined by $f(x) = \sqrt{|x|}$ for all $x \in \mathbb{R}$ we have that $f'(0) = \infty$, hence $f$ is not differentiable at $x = 0$. Similarly, the function $g$ from $\mathbb{R}$ to itself defined by $g(x) = \sqrt[3]{x}$ for all $x \in \mathbb{R}$ we have that $f'(0) = +\infty$ and neither $g$ is differentiable at $x = 0$.

© The Author(s), under exclusive license to Springer Nature Switzerland AG 2022
P. Toni et al., *100+1 Problems in Advanced Calculus*, Problem Books
in Mathematics, https://doi.org/10.1007/978-3-030-91863-7_4

*Remark* The definition of derivative given above is a bit more general than the definition typically proposed in standard textbooks, because here we do not require that the function $f$ is defined on an interval. In order to define $f'(c)$ we only need that $c$ is both a point and a limit point of the domain of $f$. This generality has some advantages. For example, it allows to include the definition of right and left derivatives in a natural way, since the right derivative $f'_+(c)$ can be defined as the derivative of the restriction of $f$ to $\mathcal{A} \cap [c, \infty[$ (provided $c$ is a limit point of $\mathcal{A} \cap [c, +\infty[)$ and $f'_-(c)$ can be defined as the derivative of the restriction of $f$ to $\mathcal{A} \cap ] - \infty, c]$ (provided $c$ is a limit point of $\mathcal{A} \cap ] - \infty, c]$). Recall that the standard definitions of right and left derivatives read as follows:

$$f'_+(c) = \lim_{x \to c^+} \frac{f(x) - f(c)}{x - c}, \quad \text{and} \quad f'_-(c) = \lim_{x \to c^-} \frac{f(x) - f(c)}{x - c}.$$

Moreover, we observe that the definition of derivative given above is equivalent to setting

$$f'(c) = \lim_{h \to 0} \frac{f(c + h) - f(c)}{h}$$

where one has to consider only those values of $h$, sufficiently close to zero, such that $c + h \in \mathcal{A}$ (which is straightforward in case $f$ is defined on a interval containing $c$).

One of the most important elementary theorems on differentiation is certainly the following:

**Theorem 4.2** *Let $\mathcal{A}$ be a subset of $\mathbb{R}$ and $f$ a function from $\mathcal{A}$ to $\mathbb{R}$. Assume that $c \in \mathcal{A}$ is a limit point of $\mathcal{A}$. If $f$ is differentiable at $c$ then it is also continuous at $c$.*

The proof is an easy exercise and can be found in any textbook. Thus differentiability implies continuity, while continuity does not imply differentiability as the classical example $f(x) = |x|$ shows: indeed, this function is continuous in the whole of $\mathbb{R}$ but is not differentiable at $x = 0$.

## 4.2  Problems

### 4.2.1  Problem 45

Which of the following six functions (see Fig. 4.1) is tangent in intuitive[1] sense to the $x$-axis at the point $x = 0$?

---

[1] Here one should use the notion of tangent line considered in the case of conics and try to extend it in a natural way to these curves, before the notion of derivative is introduced.

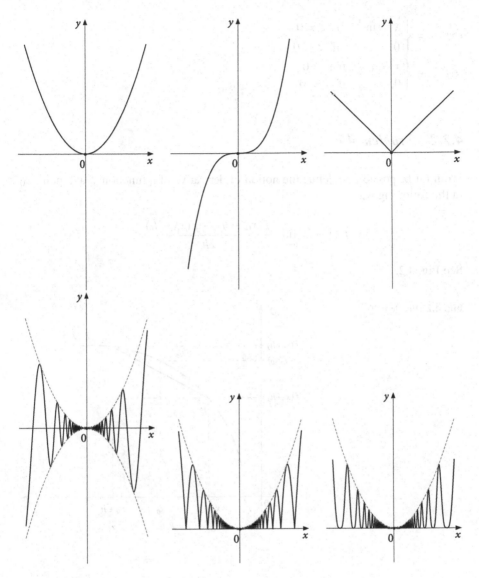

**Fig. 4.1** Problem 45

(1) $y = x^2$

(2) $y = x^3$

(3) $y = |x|$

(4) $y = \begin{cases} x^2 \sin \frac{1}{x}, & \text{if } x \neq 0 \\ 0, & \text{if } x = 0 \end{cases}$

(5) $y = \begin{cases} x^2 \left| \sin \frac{1}{x} \right|, & \text{if } x \neq 0 \\ 0, & \text{if } x = 0 \end{cases}$

(6) $y = \begin{cases} x^2 \sin^2 \frac{1}{x}, & \text{if } x \neq 0 \\ 0, & \text{if } x = 0 \end{cases}$

### 4.2.2   Problem 46

Would it be possible to define the notion of derivative of a function $f$ at a point $x_0$ in the following way

$$\widetilde{f'}(x_0) = \lim_{h \to 0} \frac{f(x_0 + h) - f(x_0 - h)}{2h} \ ?$$

See Fig. 4.2.

**Fig. 4.2** Problem 46

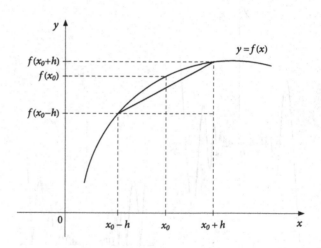

### 4.2.3   Problem 47

Would it be possible to define the notion of derivative of a function $f$ at a point $x_0$ in the following way

$$\overline{f'}(x_0) = \lim_{h \to 0} \frac{f(x_0 + 2h) - f(x_0 + h)}{h} \ ?$$

See Fig. 4.3.

**Fig. 4.3** Problem 47

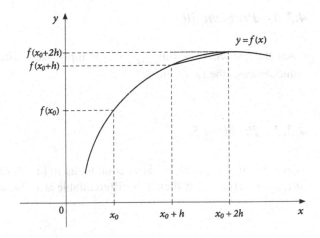

### 4.2.4  Problem 48

Let $f, g$ be functions defined in a neighbourhood of $x_0$, with $f(x) \leq g(x)$, $f(x_0) = g(x_0)$, $f'(x_0) = g'(x_0)$. Let $h$ be an arbitrary function with $f(x) \leq h(x) \leq g(x)$, see Fig. 4.4. Prove that $h$ is differentiable at $x_0$ and that $h'(x_0) = f'(x_0)$.

**Fig. 4.4** Problem 48

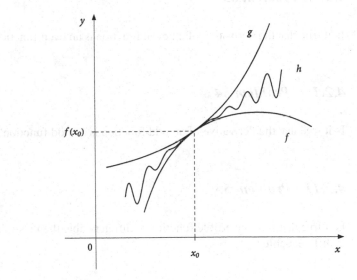

### 4.2.5  Problem 49

Give the definition of a function which is continuous and differentiable in the whole of $\mathbb{R}$, such that its derivative is continuous but not differentiable at some point.

### 4.2.6  Problem 50

Does the existence of $\lim_{x \to x_0^+} f'(x)$ imply the existence of $f'_+(x_0)$? and the coincidence of the two values?

### 4.2.7  Problem 51

Prove that if $f : [a, b] \to \mathbb{R}$ is continuous in $[a, b]$, differentiable in $(a, b)$ and $\lim_{x \to a} f'(x) = \lambda \in \mathbb{R}$ then $f$ is differentiable at $x = a$ and $f'(a) = \lambda$.

### 4.2.8  Problem 52

Does the existence of $f'(x_0)$ imply the existence of $\lim_{x \to x_0} f'(x)$?

### 4.2.9  Problem 53

Is it true that the derivative of an even function is an even function?

### 4.2.10  Problem 54

Is it true that the derivative of an odd function is an odd function?

### 4.2.11  Problem 55

Is it true that if an invertible function is differentiable then also the inverse function is differentiable?

### 4.2.12  Problem 56

The function $y = \cos x$ has been inverted by mathematicians of former times in the interval $[0, \pi]$. Choosing another interval would have had only disadvantages or also some advantages that were overlooked?

## 4.3  Solutions

### 4.3.1  Solution 45

If the word 'intuitive' leads us to the idea of a line 'caressing' the curve at the point under consideration, then it could be natural to say that the curve is tangent the $x$-axis at the point zero in the cases (1), (2) and possibly (6).

However, at a formal level, the notion of tangent line is linked to the notion of derivative by means of the relation

$$m_{\text{tangent line}} = f'(x_0)$$

hence it is necessary to consider also the cases (4) and (5).

In particular, we note that the curve in (5) probably doesn't fit a natural and spontaneous sense of tangency, and shows that the definition of derivative widens the classical point of view.

Strangely enough, there is less resistance in accepting the fact that a 'fog function' randomly generated as follows

$$y = \begin{cases} \text{random } (x), & \text{with } 0 < \text{random}(x) < x^2, \text{ if } x \neq 0 \\ 0, & \text{if } x = 0 \end{cases}$$

is tangent the $x$-axis at the origin, in which case one may imagine a kind of a 'puff' rather than a 'caress' (see Fig. 4.5).

**Fig. 4.5** Solution 45

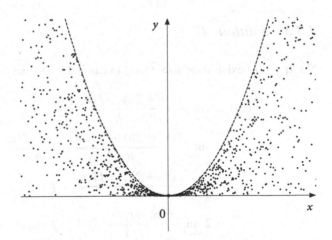

### 4.3.2   Solution 46

No. If $f'(x_0)$ exists then also $\widetilde{f}'(x_0)$ exists and the values coincide:

$$
\begin{aligned}
\widetilde{f}'(x_0) &= \lim_{h \to 0} \frac{f(x_0 + h) - f(x_0 - h)}{2h} \\
&= \lim_{h \to 0} \frac{f(x_0 + h) - f(x_0) + f(x_0) - f(x_0 - h)}{2h} \\
&= \frac{1}{2} \lim_{h \to 0} \frac{f(x_0 + h) - f(x_0)}{h} + \frac{1}{2} \lim_{h \to 0} \frac{f(x_0) - f(x_0 - h)}{h} \\
&= \frac{1}{2} f'(x_0) + \frac{1}{2} f'(x_0) = f'(x_0)
\end{aligned}
\tag{4.2}
$$

However $\widetilde{f}'(x_0)$ may exist but not $f'(x_0)$.

*Example*

$$
y = |x|
$$
$$
\widetilde{f}'(0) = \lim_{h \to 0} \frac{f(h) - f(-h)}{2h} = \lim_{h \to 0} \frac{|h| - |-h|}{2h} = 0
$$

but $f'(0)$ does not exist.

### 4.3.3   Solution 47

No. If $f'(x_0)$ exists then also $\overline{f}'(x_0)$ exists and the values coincide:

$$
\begin{aligned}
\overline{f}'(x_0) &= \lim_{h \to 0} \frac{f(x_0 + 2h) - f(x_0 + h)}{h} \\
&= \lim_{h \to 0} \frac{f(x_0 + 2h) - f(x_0)}{h} - \lim_{h \to 0} \frac{f(x_0 + h) - f(x_0)}{h} \\
&= \lim_{h \to 0} \frac{f(x_0 + 2h) - f(x_0)}{h} - f'(x_0) \\
&= 2 \lim_{h \to 0} \frac{f(x_0 + t) - f(x_0)}{t} - f'(x_0) = f'(x_0),
\end{aligned}
$$

where we have set $2h = t$. However $\overline{f}'(x_0)$ may exist but not $f'(x_0)$.

*Example*

$$y = \begin{cases} x, & \text{if } x \neq 1 \\ 2, & \text{if } x = 1 \end{cases}$$

In this case

$$\overline{f'}(1) = \lim_{h \to 0} \frac{f(1 + 2h) - f(1 + h)}{h} = \lim_{h \to 0} \frac{1 + 2h - 1 - h}{h} = 1$$

but $f'(1)$ does not exist since $f$ is not continuous at $x = 1$. Note that the definition of $\overline{f'}(1)$ does not even require that $f$ is defined at $x = 1$.

### 4.3.4 Solution 48

Since $f(x) \leq h(x) \leq g(x)$ for all $x$ and $f(x_0) = h(x_0) = g(x_0)$, we have

$$f(x_0 + k) \leq h(x_0 + k) \leq g(x_0 + k)$$

subtracting to each term of the inequality the same number (Theorem $T_1$)

$$f(x_0 + k) - f(x_0) \leq h(x_0 + k) - h(x_0) \leq g(x_0 + k) - g(x_0)$$

multiplying each term of the inequality by $\frac{1}{k}$ with $k > 0$ (Theorem $T_3$)

$$\frac{f(x_0 + k) - f(x_0)}{k} \leq \frac{h(x_0 + k) - h(x_0)}{k} \leq \frac{g(x_0 + k) - g(x_0)}{k}$$

Since $\lim_{k \to 0^+} \frac{f(x_0+k)-f(x_0)}{k} = f'(x_0)$ and $\lim_{k \to 0^+} \frac{g(x_0+k)-g(x_0)}{k} = g'(x_0)$ and $f'(x_0) = g'(x_0)$ we obtain that $\lim_{k \to 0^+} \frac{h(x_0+k)-h(x_0)}{k} = f'(x_0)$ by applying the Squeeze Theorem.

$$\frac{f(x_0+k)-f(x_0)}{k} \leq \frac{h(x_0+k)-h(x_0)}{k} \leq \frac{g(x_0+k)-g(x_0)}{k}$$

$$f'(x_0) = g'(x_0) \quad \text{as } k \to 0^+$$

In the same way we can prove that

$$\lim_{k \to 0^-} \frac{h(x_0+k)-h(x_0)}{k} = f'(x_0)$$

hence

$$\lim_{k \to 0} \frac{h(x_0+k)-h(x_0)}{k} = f'(x_0)$$

which means that $h$ is differentiable at $x_0$ and its derivative equals $f'(x_0)$.

### 4.3.5   Solution 49

*Example* See Figs. 4.6, 4.7 and 4.8.

$$y = \begin{cases} x^2, & \text{if } x > 0 \\ 0, & \text{if } x = 0 \\ -x^2, & \text{if } x < 0 \end{cases}$$

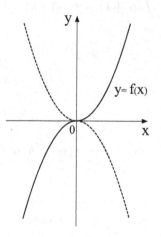

**Fig. 4.6** Solution 49

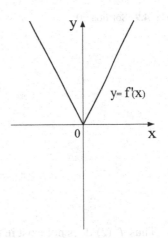

$$y' = \begin{cases} 2x, & \text{if } x > 0 \\ 0, & \text{if } x = 0 \\ -2x, & \text{if } x < 0 \end{cases}$$

**Fig. 4.7** Solution 49

$$y'' = \begin{cases} 2, & \text{if } x > 0 \\ \text{not defined}, & \text{if } x = 0 \\ -2, & \text{if } x < 0 \end{cases}$$

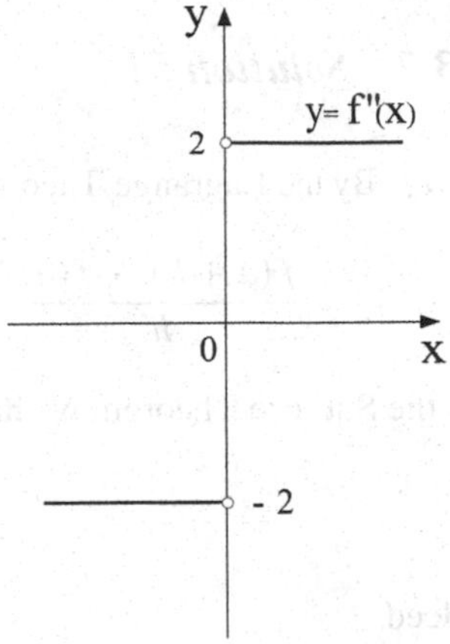

**Fig. 4.8** Solution 49

### 4.3.6 Solution 50

*Example*

$$y = \begin{cases} x, \text{if } x > 2 \\ 0, \text{if } x \leq 2 \end{cases} \qquad y' = \begin{cases} 1, \text{if } x > 2 \\ 0, \text{if } x < 2 \end{cases}$$

hence $\lim_{x \to 2^+} y' = 1$, but

$$\lim_{h \to 0^+} \frac{f(2+h) - f(2)}{h} = \lim_{h \to 0^+} \frac{2+h-0}{h} = +\infty.$$

**Fig. 4.9** Solution 50

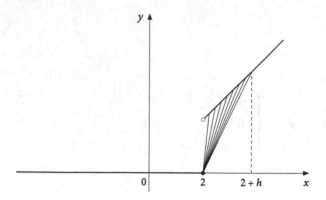

Thus $f'_+(2)$ does not exist in this case (see Fig. 4.9).

### 4.3.7   Solution 51

**Proof** By the Lagrange Theorem we have that

$$\frac{f(a+h) - f(a)}{h} = f'(\xi(h)), \quad \text{where } a \leq \xi(h) \leq a + h.$$

By the Squeeze Theorem we have that

$$\lim_{h \to 0} \xi(h) = a.$$

Indeed

$$a \quad \leq \quad \xi(h) \quad \leq \quad a + h$$

$$\searrow \quad \vdots \quad \swarrow \qquad \text{as } h \to 0$$

$$a$$

Conclusion:

$$\lim_{h \to 0} \frac{f(a+h) - f(a)}{h} = \lim_{h \to 0} f'(\xi(h)) \overset{(1)}{=} \lim_{x \to a} f'(x) \overset{(2)}{=} \lambda,$$

that is $f'(a) = \lambda$.

(1) recall that if $h \to 0$ then $\xi(h) \to a$
(2) by assumptions

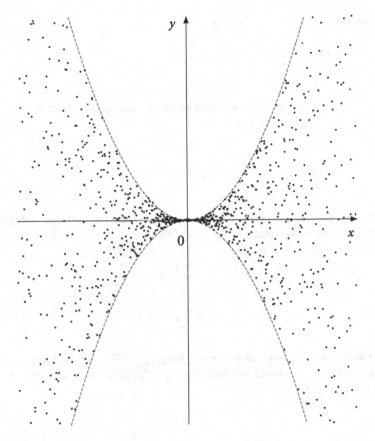

**Fig. 4.10** Solution 52

## 4.3.8 Solution 52

No.

*Example 1*
We consider the 'fog function' (see Fig. 4.10)

$$f(x) = \begin{cases} \text{random } (x) \text{ with } 0 < \text{random } (x) \leq x^2, & \text{if } x \in \mathbb{Q} \\ 0, & \text{if } x = 0 \\ \text{random } (x) \text{ with } -x^2 \leq \text{random } (x) < 0, & \text{if } x \notin \mathbb{Q} \end{cases}$$

Recall that here random $(x)$ means a random number depending on $x$.
Let us compute the difference quotient of $f$ at the point $x = 0$:

$$\frac{f(0+h) - f(0)}{h} = \frac{f(h) - 0}{h} = \frac{f(h)}{h}.$$

Now, since $-x^2 \le f(x) \le x^2$, we have:

$$-h^2 \le f(h) \le h^2$$

> multiplying each term of the inequality by $\frac{1}{h}$ with $h > 0$
> (Theorem $T_3$)

$$-h \le \frac{f(h)}{h} \le h$$

Thus, considering that $\lim_{h \to 0^+} h = \lim_{h \to 0^+}(-h) = 0$ and applying the Squeeze Theorem yield $\lim_{h \to 0^+} \frac{f(h)}{h} = 0$. Indeed

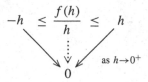

In an analogous way, we can prove that $\lim_{h \to 0^-} \frac{f(h)}{h} = 0$ hence $\lim_{h \to 0} \frac{f(h)}{h} = 0$, that is $f'(0) = 0$. However, the limit $\lim_{x \to 0} f'(x)$ does not exist since not even $f'(x)$ exists if $x \ne 0$.

*Example 2*

$$f(x) = \begin{cases} x^2 \sin \frac{1}{x}, & \text{if } x \ne 0 \\ 0, & \text{if } x = 0 \end{cases}$$

Even in this case we have that $f'(0) = 0$ (it suffices to note that $-1 \le \sin \frac{1}{x} \le 1$ hence $-x^2 \le x^2 \sin \frac{1}{x} \le x^2$ and proceed as in the previous example). However, the limit $\lim_{x \to 0} f'(x) = \lim_{x \to 0}(2x \sin \frac{1}{x} - \cos \frac{1}{x})$ does not exist.

*Intuitive example*
We define a function step by step with the following procedure.

*Step 1.*   We consider the lines $y = -x + \frac{1}{2^n}$ with $n \in \mathbb{N}$ and $0 \le x \le 1$ (see Fig. 4.11).

*Step 2.*   We draw the line $y = x$ and the parabola $y = x^2$, and we recall that if $0 < x < 1$ then $x^2 < x$ (see Fig. 4.11).

*Step 3.*   We consider the segments defined on the lines $y = -x + \frac{1}{2^n}$ by the intersections of those lines with the parabola $y = x^2$ and the line $y = 0$ (see Fig. 4.12).

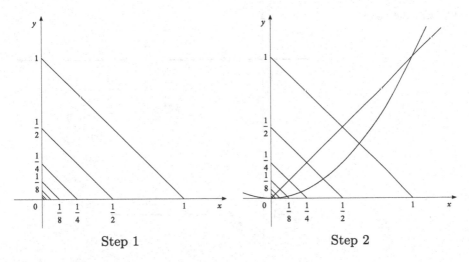

**Fig. 4.11** Solution 52

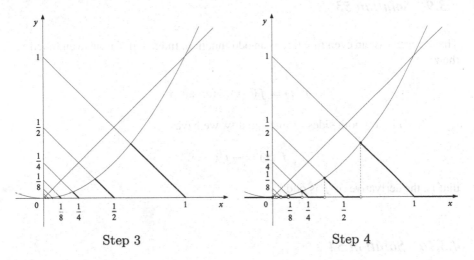

**Fig. 4.12** Solution 52

*Step 4.* We consider the function defined the graph of which is the union of the segments found in Step 3 where such segments are defined, and which equals zero elsewhere (see Fig. 4.12).

The function defined above is differentiable at zero ($f'(0) = 0$: to prove it, just compare $f$ with the functions $y = 0$ and $y = x^2$). However, it is absolutely evident that $\lim_{x \to 0} f'(x)$ does not exist: it suffices to draw the graph of $f'(x)$. See Fig. 4.13.

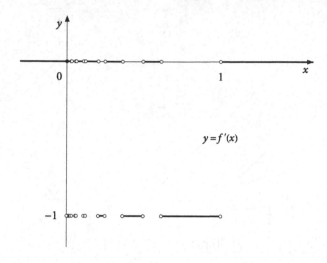

**Fig. 4.13** Solution 52

## 4.3.9   Solution 53

The derivative of an even function is an odd function. Indeed, if $f$ is an even function then

$$f(x) = f(-x), \quad \text{for all } x.$$

By differentiation both sides of this equality, we have

$$f'(x) = -f'(-x),$$

that is, the derivative of $f$ is odd.

## 4.3.10   Solution 54

The derivative of an odd function is an even function. Indeed, if $f$ is an odd function then

$$f(x) = -f(-x), \quad \text{for all } x.$$

By differentiation both sides of this equality, we have

$$f'(x) = -(-f'(-x)) = f'(-x),$$

that is, the derivative of $f$ is even.

**Fig. 4.14** Solution 55

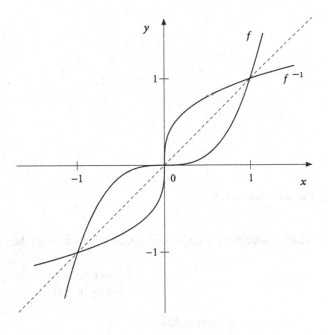

### 4.3.11 Solution 55

In general no.

*Example* See Fig. 4.14.

$\sqrt[3]{x}$ is not differentiable at $x = 0$ because:

$$\lim_{h \to 0} \frac{f^{-1}(0+h) - f^{-1}(0)}{h} = \lim_{h \to 0} \frac{\sqrt[3]{h}}{h} = +\infty.$$

### 4.3.12 Solution 56

If they would have chosen the interval $[-\pi, 0]$, overcoming the natural aversion to negative numbers (usually placed at the left-hand of zero, hence a bit sinister) they would have obtained the function

$$\begin{cases} y = \arccos x \\ -\pi \le y \le 0 \\ -1 \le x \le 1 \end{cases}$$

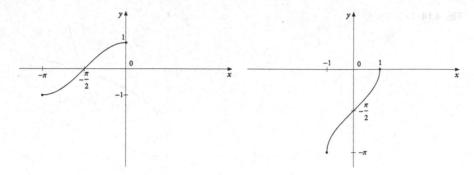

**Fig. 4.15** Solution 56

which would have been increasing, as it is increasing the function

$$\begin{cases} y = \cos x \\ -\pi \le x \le 0 \end{cases}$$

with the extra advantage that:

$$(\arccos x)' = (\arcsin x)' = \frac{1}{\sqrt{1 - x^2}},$$

that is, one single formula for the derivative of both functions $\arccos x$ and $\arcsin x$. See Fig. 4.15.

We describe now three different ways of proving that with the new definition we would get

$$(\arccos x)' = (\arcsin x)' = \frac{1}{\sqrt{1 - x^2}}.$$

**1st WAY**

It suffices to observe that the graph of the cosine function, considered in the interval $[-\pi, 0]$, can be obtained by translating by $\pi/2$ the graph of the sine function considered in the interval $[-\pi/2, \pi/2]$ (which is the interval where the sine function is usually inverted).

As it is evident from Fig. 4.16, we have that:

$$\arccos a = \arcsin a - \frac{\pi}{2}, \quad \text{with } -1 \le a \le 1$$

and, changing names:

$$\arccos x = \arcsin x - \frac{\pi}{2}, \quad \text{with } -1 \le x \le 1.$$

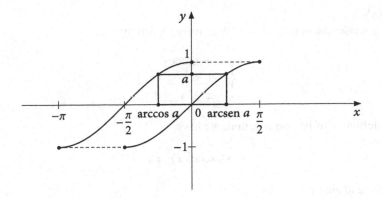

**Fig. 4.16** Solution 56

Thus, differentiating both sides we get

$$D(\arccos x) = D(\arcsin x) - D\left(\frac{\pi}{2}\right)$$

$$= D(\arcsin x) - 0 = D(\arcsin x) = \frac{1}{\sqrt{1-x^2}}.$$

**2nd WAY**
By restricting the cosine function to the interval $[-\pi, 0]$

$$\begin{cases} y = \cos x \\ -\pi \leq x \leq 0 \end{cases}$$

and inverting it, we obtain

$$\begin{cases} x = \arccos y \\ -1 \leq y \leq 1 \\ -\pi \leq x \leq 0 \end{cases} \rightarrow \sin x < 0 \rightarrow -\sin x > 0$$

Thus, differentiating the function $x = \arccos y$ we get:

$$x' = \frac{dx}{dy} = \frac{1}{\frac{dy}{dx}} = \frac{1}{-\sin x} = \frac{1}{\sqrt{1-\cos^2 x}} = \frac{1}{\sqrt{1-y^2}}$$

and, swapping the letters $x$ and $y$ we eventually get that

$$\text{if } y = \arccos x \text{ then } y' = \frac{1}{\sqrt{1-x^2}}.$$

**3rd WAY**

Let us consider the new version of the inverse function:

$$\begin{cases} y = \arccos x \\ -1 \le x \le 1 \\ -\pi \le y \le 0 \end{cases}$$

By definition of inverse function, we have

$$\cos(\arccos x) = x$$

and differentiating both sides

$$-\sin(\arccos x) \cdot y' = 1$$

hence

$$y' = \frac{1}{-\sin(\arccos x)} = \frac{1}{-(-\sqrt{1-x^2})} = \frac{1}{\sqrt{1-x^2}}.$$

# Chapter 5
# Classical Theorems of differential calculus

## 5.1 Theoretical background

In this chapter, we consider the classical theorems of differential calculus such as Rolle's, Lagrange's, Cauchy's and De L'Hospital's theorems. We recall here their statements and we give some hints about their proofs.

### 5.1.1 Rolle's Theorem

**Theorem 5.1** *Let $f$ be a function from $[a, b]$ to $\mathbb{R}$ with $a, b \in \mathbb{R}$, $a < b$, such that*

  (i) *$f$ is continuous on $[a, b]$;*
 (ii) *$f$ is differentiable on $(a, b)$;*
(iii) *$f(a) = f(b)$.*

*Then, there exists $c \in (a, b)$ such that $f'(c) = 0$.*

### 5.1.2 Lagrange's Theorem

**Theorem 5.2** *Let $f$ be a function from $[a, b]$ to $\mathbb{R}$ with $a, b \in \mathbb{R}$, $a < b$, such that*

  (i) *$f$ is continuous on $[a, b]$;*
 (ii) *$f$ is differentiable on $(a, b)$.*

© The Author(s), under exclusive license to Springer Nature Switzerland AG 2022
P. Toni et al., *100+1 Problems in Advanced Calculus*, Problem Books
in Mathematics, https://doi.org/10.1007/978-3-030-91863-7_5

*Then, there exists $c \in (a, b)$ such that*

$$f'(c) = \frac{f(b) - f(a)}{b - a}.$$

### 5.1.3   Cauchy's Theorem

**Theorem 5.3** *Let $f$ and $g$ be two functions defined from $[a, b]$ to $\mathbb{R}$ with $a, b \in \mathbb{R}$, $a < b$, such that $f, g$ are continuous on $[a, b]$ and differentiable on $(a, b)$. Then there exists $c \in (a, b)$ such that*

$$(f(b) - f(a))g'(c) = (g(b) - g(a))f'(c).$$

The following theorem is stated for arbitrary open intervals $I$ in $\mathbb{R}$, bounded or unbounded. This means that $I$ can be any interval of the type $(a, b)$, $(a, +\infty[$, $(-\infty, b)$, $(-\infty, +\infty)$. As usual, $a, b$ are called end points. This classical terminology is used here also in the case of unbounded intervals, in which case $+\infty$ and $-\infty$ will also be considered end points.

### 5.1.4   De L'Hospital Theorem

**Theorem 5.4** *Let $I$ be an open interval in $\mathbb{R}$ and let $\square$ be an endpoint of $I$. Let $f, g$ be two differentiable functions from $I$ to $\mathbb{R}$. Assume that the derivative of $g$ never vanishes and that the limit*

$$\lim_{x \to \square} \frac{f'(x)}{g'(x)} = \Delta$$

*exist, with $\Delta$ being either a real number, or $+\infty$, $-\infty$. Assume also that*

$$\lim_{x \to \square} f(x) = \lim_{x \to \square} g(x) = 0,$$

*or*

$$\lim_{x \to \square} g(x) = +\infty,$$

*or*

$$\lim_{x \to \square} g(x) = -\infty.$$

*Then*

$$\lim_{x \to \square} \frac{f(x)}{g(x)} = \triangle.$$

Note that De L'Hospital Theorem is often applied in order to calculate the limit of functions in the case of the classical indeterminate forms

$$\frac{0}{0} \quad \text{and} \quad \frac{\pm\infty}{\pm\infty}.$$

However, its proof does not requite that the function $f$ at the numerator diverges to infinity in the second and third case (hence it is not required to assume it).

The rigorous proof of these theorems can be found in almost all textbooks. Here we shall give some hints on their geometric interpretation. To begin with, we would like emphasize the fact that, once Rolle's Theorem is stated, then the proof of Lagrange's Theorem immediately follows, even at an intuitive level. This is explained in what follows.

### 5.1.5 Geometric interpretation of the proof of Lagrange's Theorem

In order to prove Lagrange's Theorem (see Fig. 5.1), it suffices to find a function that 'Rollefies' Lagrange's Theorem (that is, a function that reduces Lagrange's

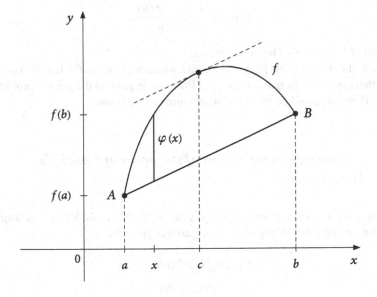

**Fig. 5.1** Geometric interpretation of Lagrange's Theorem

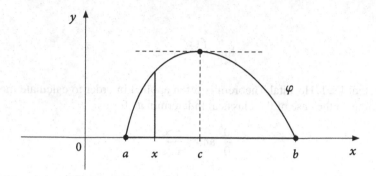

**Fig. 5.2**  Reduction to Rolle's Theorem

Theorem to Rolle's Theorem). Namely, it is convenient to set

$$\varphi(x) = y_{\text{curve}} - y_{\text{line AB}}$$

that is

$$\varphi(x) = f(x) - \left( \frac{f(b) - f(a)}{b - a}(x - a) + f(a) \right) .$$

Function $\varphi$ fulfils the hypothesis of Rolle's Theorem, hence there exists $c \in (a, b)$ such that $\varphi'(c) = 0$, that is $\varphi'(c) = f'(c) - m_{AB} = 0$ hence $f'(c) = m_{AB}$. Since

$$m_{AB} = \frac{f(b) - f(a)}{b - a},$$

the proof of Lagrange's Theorem follows.

Note that in Fig. 5.2, the point $c$ is a point which maximizes the function $\varphi$, which means that $\varphi$ reaches its maximum at $c$. However, in general the point $c$ provided by Rolle's Theorem could be any local maximum or minimum.

## 5.1.6   Geometric interpretation of the proof of Cauchy's Theorem

For simplicity we assume that $f(a) = g(a) = 0$. We consider the rectangles in Fig. 5.3 with sides depending on $x$. Their area is given by

$$S_f(x) = f(x)(g(b) - g(a))$$
$$S_g(x) = g(x)(f(b) - f(a))$$

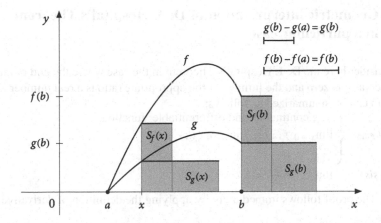

**Fig. 5.3** Geometric interpretation of Cauchy's Theorem

We have that $S_f(a) = S_g(a) = 0$ and $S_f(b) = S_g(b)$. Thus, we can apply Rolle's Theorem to the function

$$S_f(x) - S_g(x)$$

hence there exists $c \in (a, b)$ such that

$$S'_f(c) - S'_g(c) = 0$$

that is

$$f'(c)(g(b) - g(a)) = g'(c)(f(b) - f(a)).$$

Thus, keeping in mind the fact that the point $c$ provided by Rolle's Theorem is actually a point which maximizes or minimizes the function $S_f(x) - S_g(x)$, we understand that the point $c$ from Cauchy's Theorem is related to the gap between the size of two rectangles and 'appears' when the gap reaches a global maximum or a minimum (but also when it reaches a local maximum or minimum).

*Remark* Is there a loss of generality in assuming that $f(a) = g(a) = 0$? No. If the condition $f(a) = g(a) = 0$ is not satisfied then we can simply consider the functions $g_1(x) = g(x) - g(a)$ and $f_1(x) = f(x) - f(a)$. Indeed:

$$g'_1(x) = g'(x), \quad f'_1(x) = f(x), \quad g_1(b) - g_1(a) = g(b) - g(a)$$

$$f_1(b) - f_1(a) = f(b) - f(a), \quad g_1(a) = f_1(a) = 0.$$

## 5.2   Geometric interpretation of De L'Hospital's Theorem in a particular case

We consider here the De L'Hospital's Theorem in the case where the end point $\square$ of the interval $I$ is zero and the limit $\triangle$ of the appropriate ratio is a real number $\lambda$. The theorem can be summarized as follows:

$$Hypothesis \begin{cases} f, g \text{ continuous and differentiable functions} \\ \lim_{x \to 0} f(x) = \lim_{x \to 0} g(x) = 0 \\ \frac{f'(0)}{g'(0)} = \lambda \end{cases}$$

$$Thesis \quad \lim_{x \to 0} \frac{f(x)}{g(x)} = \lambda$$

**Proof** The proof follows immediately by applying the definition of derivative:

$$\lim_{x \to 0} \frac{f(x)}{g(x)} \overset{(1)}{=} \lim_{x \to 0} \frac{f(x) - f(0)}{g(x) - g(0)} = \lim_{x \to 0} \frac{\frac{f(x) - f(0)}{x - 0}}{\frac{g(x) - g(0)}{x - 0}} \overset{(2)}{=} \frac{f'(0)}{g'(0)} \overset{(3)}{=} \lambda$$

Note that each equality is justified as follows:

(1)  Since $f$ and $g$ are continuous and $\lim_{x \to 0} f(x) = \lim_{x \to 0} g(x) = 0$ then $f(0) = g(0) = 0$ hence $f(x) = f(x) - f(0)$ and $g(x) = g(x) - g(0)$.
(2)  By definition of derivative and the theorem on the limit of a quotient.
(3)  By assumptions.

By inspecting this proof and comparing it with Fig. 5.4, it appears that the ratio between $f(x)$ and $g(x)$ is nothing but the ratio between the tangents of the two angles indicated in Fig. 5.4. Then, it is clear that passing to the limit as $x \to 0$ boils down to considering the ratio between the slopes of the two tangent lines of the graphs of $f$ and $g$ at the point $x = 0$.

**Fig. 5.4** Geometric interpretation

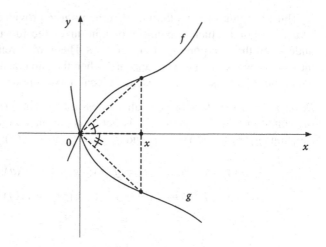

## 5.3  Problems

### 5.3.1  Problem 57

Among the assumptions of Rolle's Theorem, is it possible to drop the condition that
the function defined on an interval $[a, b]$ is continuous at $a$ and $b$?

### 5.3.2  Problem 58

If a function fulfils all assumptions of Rolle's Theorem but is not differentiable at a
point, does the theorem still apply to it?

### 5.3.3  Problem 59

Given the function $f(x) = |x^2 - 4x|$, find the values of $a \in \mathbb{R}$ such that the
assumptions of Rolle's Theorem are fulfilled for $f$ on the interval $[a, a + 2]$.

### 5.3.4  Problem 60

Prove that the following variant of Rolle's Theorem holds: let $f$ be a func-
tion defined on $(a, b)$, continuous, differentiable, and such that $\lim_{x \to a} f(x) = \lim_{x \to b} f(x)$. Then there exists $c \in (a, b)$ such that $f'(c) = 0$.

### 5.3.5  Problem 61

Let $f$ be a function defined on $\mathbb{R}$, continuous, differentiable and such that
$\lim_{x \to +\infty} f(x) = \lim_{x \to -\infty} f(x)$. Is there $c \in \mathbb{R}$ such that $f'(c) = 0$?

### 5.3.6  Problem 62

Given the function $f(x) = |1/x|$, find the values of $a \in \mathbb{R}$ such that the assumptions
of Lagrange's Theorem are fulfilled for $f$ on the interval $[a, a + 3]$.

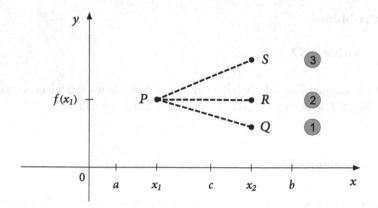

**Fig. 5.5** Problem 63.

### 5.3.7  Problem 63

Let $f$ be a real-valued function defined on the interval $(a, b)$. Assume that $P = (x_1, f(x_1))$ is a point of the graph of $f$. Let $x_2 \in (a, b)$ with $x_2 > x_1$. Let $Q = (x_2, y_Q)$, $R = (x_2, y_R)$, $S = (x_2, y_S)$ be three points with abscissa equal to $x_2$, such that

$$y_Q < y_R = f(x_1) < y_S,$$

see Fig. 5.5. Assume that $f$ is differentiable and that $f'(x) > 0$ for all $x \in (a, b)$. Which one of the following three cases

(1)  $Q$ belongs to the graph of $f$
(2)  $R$ belongs to the graph of $f$
(3)  $S$ belongs to the graph of $f$

is compatible with Lagrange's Theorem?

### 5.3.8  Problem 64

Is it true or false that De L'Hospital Theorem allows to claim that

$$\text{If} \quad \begin{cases} \lim_{x \to \square} f(x) = \lim_{x \to \square} g(x) = \Delta \quad \text{with} \quad \Delta = 0, +\infty, -\infty \\ \lim_{x \to \square} \frac{f'(x)}{g'(x)} \text{ doesn't exist} \end{cases}$$

then $\lim_{x \to \square} \frac{f(x)}{g(x)}$ doesn't exist?

## 5.4  Solutions

### 5.4.1  Solution 57

No. Indeed, consider the function $f$ defined as follows (see Fig. 5.6):

$$f(x) = \begin{cases} x, & \text{if } 0 < x \le 1 \\ 1, & \text{if } x = 0 \end{cases}$$

Function $f$ satisfies all the hypotheses of Rolle's Theorem with the exception of the continuity at 0, but for all $c \in (0, 1)$ we have $f'(c) = 1 \ne 0$.

**Fig. 5.6**  Solution 57

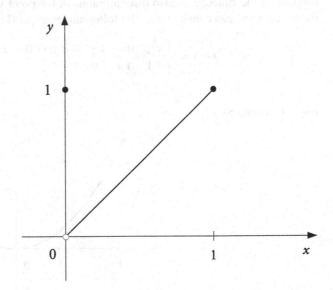

### 5.4.2  Solution 58

No.

*Example*

$$f(x) = |x| \quad \text{with} -1 \le x \le 1$$

See Fig. 5.7.

Function $f$ satisfies all the hypotheses of Rolle's Theorem with the exception of the differentiability at 0, but for all $c \in (-1, 1) \setminus \{0\}$ we have $f'(c) \ne 0$.

**Fig. 5.7** Solution 58

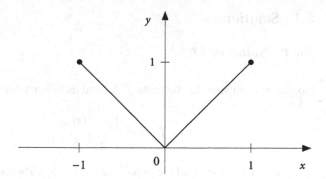

*Remark* If the function is also discontinuous at the point where it is not differentiable, then one can consider also the following example (Fig. 5.8):

$$f(x) = \begin{cases} x^2, & \text{if } -1 \le x < 0 \text{ or } 0 < x \le 1 \\ -1, & \text{if } x = 0 \end{cases}$$

**Fig. 5.8** Solution 58

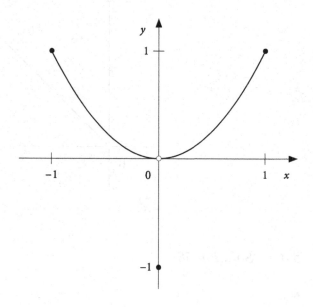

### 5.4.3  Solution 59

$a = 1$. Indeed:

$$f(x) = \begin{cases} x^2 - 4x, & \text{if } x \le 0 \text{ or } x \ge 4 \\ -x^2 + 4x, & \text{if } 0 < x < 4 \end{cases}$$

**Fig. 5.9** Solution 59

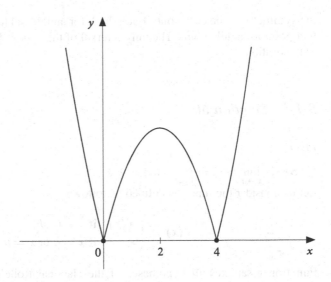

See Fig. 5.9

The problem can be solved in two ways: one analytic, another one synthetic.

**1st Method**

We analyse all possible cases:

If $a \leq -2$ then the interval $[a, a+2]$ is contained in the half-line $(-\infty, 0]$ where $f$
   is strictly decreasing, hence the condition $f(a) = f(a+2)$ is never satisfied.

If $-2 < a < 0$ then the interval $(a, a+2)$ always contains the point $x = 0$ where
   $f$ is not differentiable.

If $0 \leq a \leq 2$ then the interval $[a, a+2]$ is contained in the interval $[0, 4]$ where
   the hypothesis of continuity and differentiability required by Rolle's Theorem
   are satisfied. Moreover, since $f$ is symmetric with respect to the line $x = 2$, it is
   immediate to see that the only value of $a$ for which $f(a) = f(a+2)$ is $a = 1$.

If $2 < a < 4$ then the interval $(a, a+2)$ always contains the point $x = 4$ where $f$
   is not differentiable.

If $a \geq 4$ then the interval $[a, a+2]$ is contained in the half-line $[4, +\infty)$ where
   the function is strictly increasing, hence the condition $f(a) = f(a+2)$ is never
   satisfied.

*Conclusion* The only value of $a$ such that function $f$ is continuous in $[a, a+2]$,
differentiable in $(a, a+2)$ and $f(a) = f(a+2)$ is $a = 1$.

**2nd Method**

We exploit the thesis of the theorem and the symmetry of the curve, by arguing
as follows. If the hypothesis of the theorem holds true then also the thesis of the
theorem holds true: thus, in the interval $[a, a+2]$ there is a point where the derivative
of $f$ vanishes. The only point where the derivative of $f$ vanishes is $x = 2$ and by

the symmetry of the curve one should look for an interval $[a, a+2]$ which contains that point as middle point. The only interval of this type is $[1, 3]$. Thus $a = 1$ is the only solution.

### 5.4.4   Solution 60

*Proof*

**CASE 1** $\lim_{x \to a} f(x) = \lim_{x \to b} f(x) = \lambda \in \mathbb{R}$
Let us consider the function defined as follows

$$g(x) = \begin{cases} f(x), & \text{if } x \in (a, b) \\ \lambda, & \text{if } x = a \text{ or } x = b \end{cases}$$

Function $g$ satisfies all hypotheses of the classical Rolle's Theorem hence there exists $c \in (a, b)$ such that $g'(c) = f'(c) = 0$.

**CASE 2** $\lim_{x \to a} f(x) = \lim_{x \to b} f(x) = +\infty$ or $\lim_{x \to a} f(x) = \lim_{x \to b} f(x) = -\infty$
This case is left to the reader.

### 5.4.5   Solution 61

Yes.

*Proof* We consider the case $\lim_{x \to +\infty} f(x) = \lim_{x \to -\infty} f(x) = +\infty$, see Fig. 5.10.

Let $x_1 \in \mathbb{R}$ be a fixed point. If $f$ has a minimum at $x_1$ then $f'(x_1) = 0$ since $f$ is differentiable, hence the theorem is proved. Otherwise there exists $x_2 \in \mathbb{R}$ such that $f(x_2) < f(x_1)$. We assume that $x_2$ is at the right hand of $x_1$, that is $x_2 > x_1$ (we would proceed in a similar way if $x_2 < x_1$). Since $\lim_{x \to +\infty} f(x) = +\infty$ then there exists $x_3 \in \mathbb{R}$ with $x_3 > x_2$ such that $f(x_3) > f(x_1)$ (if such a point $x_3$ would not exist then we would have $f(x) \leq f(x_1)$ for all $x > x_2$, hence the function would be bounded from above and this would contradict the fact that $\lim_{x \to +\infty} f(x) = +\infty$). Since the graph of the function is below the line $r : y = f(x_1)$ at $x_2$ and is above at $x_3$, then by the hypothesis of continuity there exists $x_4$ where the graph intersects the line $r$, that is $f(x_4) = f(x_1)$. Thus the function satisfies the hypothesis of Rolles's Theorem on the interval $[x_1, x_4]$, hence there exists $c \in (x_1, x_4)$ such that $f'(c) = 0$.

In the cases $\lim_{x \to +\infty} f(x) = \lim_{x \to -\infty} f(x) = -\infty$ and $\lim_{x \to +\infty} f(x) = \lim_{x \to -\infty} f(x) = l$, we proceed in a similar way.

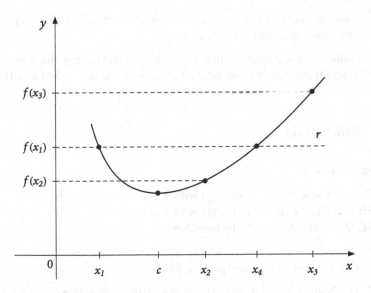

**Fig. 5.10** Solution 61

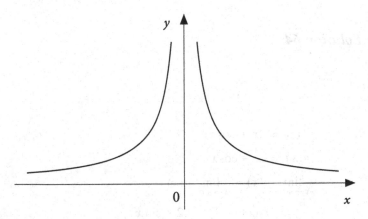

**Fig. 5.11** Solution 62

### 5.4.6 Solution 62

$a < -3$ or $a > 0$. See Fig. 5.11.

If $a < -3$ then the interval $[a, a + 3]$ is contained in the half-line $(-\infty, 0)$ where
$f$ is defined, continuous and differentiable.

If $-3 \leq a \leq 0$ then the interval $[a, a + 3]$ always contains the point $x = 0$ where $f$
is not defined.

If $a > 0$ then the interval $[a, a + 3]$ is contained in the half-line $(0, +\infty)$ where $f$ is defined, continuous and differentiable.

Thus, the values of $a$ such that the function is defined and continuous in the interval $[a, a + 3]$ and differentiable in the interval $(a, a + 3)$ are: $a < -3$ or $a > 0$.

### 5.4.7 Solution 63

The case (3). Indeed:

CASE (1): There exists $c \in (x_1, x_2)$ with $f'(c) = m_{PQ} < 0$ ⌇
CASE (2): There exists $c \in (x_1, x_2)$ with $f'(c) = m_{PR} = 0$ ⌇
CASE (3): Compatible with the hypothesis.

*Conclusion*
$f' > 0$ on $(a, b) \to f$ is increasing on $(a, b)$.

*Remark* The symbol ⌇ means 'in contradiction with the hypothesis'.

### 5.4.8 Solution 64

False

*Example*

$$f(x) = 2x + \cos x$$

$$g(x) = 2x - \cos x$$

$$\lim_{x \to +\infty} f(x) = \lim_{x \to +\infty} g(x) = +\infty$$

$$\lim_{x \to +\infty} \frac{f'(x)}{g'(x)} = \lim_{x \to +\infty} \frac{2 - \sin x}{2 + \sin x} \text{ doesn't exist}$$

However

$$\lim_{x \to +\infty} \frac{f(x)}{g(x)} = \lim_{x \to +\infty} \frac{2x + \cos x}{2x - \cos x} = \lim_{x \to +\infty} \frac{1 + \frac{\cos x}{2x}}{1 - \frac{\cos x}{2x}} = 1$$

# Chapter 6
# Monotonicity, concavity, minima, maxima, inflection points

## 6.1 Theoretical background

In this section, we recall a few classical definitions and results.

### 6.1.1 Increasing and decreasing functions

**Definition 6.1** Let $\mathcal{A}$ be a subset of $\mathbb{R}$ and $f$ be a function from $\mathcal{A}$ to $\mathbb{R}$.

(i) We say that $f$ is monotone increasing if $f(x_1) \le f(x_2)$ for all $x_1, x_2 \in \mathcal{A}$ with $x_1 \le x_2$.

(ii) We say that $f$ is monotone decreasing if $f(x_1) \ge f(x_2)$ for all $x_1, x_2 \in \mathcal{A}$ with $x_1 \le x_2$.

**Theorem 6.2** *Let $I$ be an interval in $\mathbb{R}$ (possibly bounded, unbounded, open, closed, or semi-closed) and $f$ be a differentiable function from $I$ to $\mathbb{R}$. Then the following statements hold:*

*(i) $f$ is monotone increasing if and only if $f'(x) \ge 0$ for all $x \in I$.*

*(ii) $f$ is monotone decreasing if and only if $f'(x) \le 0$ for all $x \in I$.*

### 6.1.2 Minima and maxima of a function

**Definition 6.3** Let $\mathcal{A}$ be a subset of $\mathbb{R}$ and $f$ a function from $\mathcal{A}$ to $\mathbb{R}$. Let $c \in \mathcal{A}$.

(i) We say that $c$ is a point of (global) minimum for $f$ if $f(c) \le f(x)$ for all $x \in \mathcal{A}$.

(ii) We say that $c$ is a point of local minimum for $f$ if there exists an open interval
$I$ containing $c$ such that $f(c) \leq f(x)$ for all $x \in \mathcal{A} \cap I$.
(iii) We say that $c$ is a point of (global) maximum for $f$ if $f(c) \geq f(x)$ for all
$x \in \mathcal{A}$.
(iv) We say that $c$ is a point of local maximum for $f$ if there exists an open interval
$I$ containing $c$ such that $f(c) \geq f(x)$ for all $x \in \mathcal{A} \cap I$.

### 6.1.3  Concavity of a function

In order to talk about the concavity of a function, the easiest way is probably the one
based on the notion of convexity of a planar set. Recall that a subset of the Euclidean
plane is called *convex* if it contains any segment joining two points of its. Based on
this very elementary notion we can give the following classical definition.

**Definition 6.4** Let $I$ be an interval in $\mathbb{R}$ (possibly bounded, unbounded, open,
closed, or semi-closed) and $f$ be a function from $I$ to $\mathbb{R}$.

(i) We say that $f$ is concave upward if the set $\{(x, y) \in \mathbb{R}^2 : y \geq f(x)\}$ (called
'epigraph') is a convex subset of the Euclidean plane.
(ii) We say that $f$ is concave downward if the set $\{(x, y) \in \mathbb{R}^2 : y \leq f(x)\}$ (called
'subgraph') is a convex subset of the Euclidean plane.

Equivalently, a function $f$ is concave upward if for any $x_1, x_2 \in I$ with $x_1 < x_2$
and any $t \in [0, 1]$ we have

$$f(x_1 + t(x_2 - x_1)) \leq f(x_1) + t(f(x_2) - f(x_1)), \tag{6.1}$$

which means that the graph of $f$ on the interval $[x_1, x_2]$ is below the segment joining
the points $(x_1, f(x_1))$ and $(x_2, f(x_2))$ in the plane. Note that the interval $[x_1, x_2]$ is
given by all points of the form $x_1 + t(x_2 - x_1)$ for $t$ ranging from zero to one. We
note that by setting $c = x_1 + t(x_2 - x_1)$, inequality (6.1) can be written in the form

$$f(c) \leq f(x_1) + \frac{f(x_2) - f(x_1)}{x_2 - x_1}(c - x_1) \tag{6.2}$$

The case of functions which are concave downward is obtained by reversing the
inequality in (6.1).

The terminology could be sometimes a bit confusing, since many authors
(mostly, at the level of scientific literature) prefer to talk about convex and concave
functions. The reader must be aware of the fact that

*concave upward function* is synonymous with *convex function* and

> *concave downward function* is synonymous with *concave function*

The prototypes that everybody keeps in mind are the parabolas $y = x^2$ and $y = \sqrt{x}$ : this first function is concave upward, the second is concave downward. One should be aware that the given definition is very simple but sufficiently general to include some exceptional cases. For example, strictly speaking, the real-valued function $f$ defined on $[0, +\infty[$ by setting

$$f(x) = \begin{cases} \sqrt{x}, & \text{if } x > 0, \\ -1, & \text{if } x = 0 \end{cases}$$

is still concave downward, although one would not easily believe that! These peculiarities could appear only at the boundary points of the intervals under consideration, since one can actually prove that *if a function is concave upward (or downward) then it must be continuous in the interior of the interval where it is defined.* In fact, the above function $f$ is continuous on $(0, \infty)$ and discontinuous only at $x = 0$.

In the case of regular functions, the following well-known criterion is available.

**Theorem 6.5** *Let I be an open interval in $\mathbb{R}$ (possibly bounded or unbounded) and f be a twice differentiable function from I to $\mathbb{R}$ (that is, $f''(x)$ exists and is finite for all $x \in I$). Then the following statements hold:*

*(i)  f is concave upward if and only if $f''(x) \geq 0$ for all $x \in I$.*
*(ii)  f is concave downward if and only if $f''(x) \leq 0$ for all $x \in I$.*

### 6.1.4  Inflection points

**Definition 6.6**  Let $I$ be an open interval in $\mathbb{R}$ (possibly bounded or unbounded) and $f$ be a function from $I$ to $\mathbb{R}$. Let $c \in I$. We say that $c$ is an inflection point for $f$ if one of the two possibilities occur:

(i)  either $f$ is concave upward on the right of $c$ and concave downward on the left of $c$;
(ii)  or $f$ is concave upward on the left of $c$ and concave downward on the right of $c$.

Note that saying that $f$ is concave upward or downward on the right of $c$ means that the restriction of $f$ to the set $I \cap [c, +\infty[$ is concave upward or downward respectively. Similarly, saying that $f$ is concave upward or downward on the left of $c$ means that the restriction of $f$ to the set $I \cap ] -\infty, c]$ is concave upward or downward respectively.

*Remark* Usually, the definition of inflection point is applied to continuous functions which are twice differentiable for $x \neq c$. This includes the case when the function is not differentiable at $x = c$. For example, the real-valued function $g$ defined on $\mathbb{R}$ by setting $g(x) = \sqrt[3]{x}$ for all $x \in \mathbb{R}$ has an inflection point at $x = 0$ and $f'(0) = +\infty$, hence $g$ is not differentiable at $x = 0$. Strictly speaking, also the discontinuous function $h$ defined by

$$h(x) = \begin{cases} \sqrt[3]{x}, & \text{if } x \geq 0, \\ \sqrt[3]{x} - 1, & \text{if } x < 0 \end{cases}$$

has an inflection point at $x = 0$, while the function $k$ defined by

$$k(x) = \begin{cases} \frac{1}{x}, & \text{if } x \neq 0, \\ 0, & \text{if } x = 0 \end{cases}$$

doesn't have an inflection point at $x = 0$.

## 6.2   Problems

### 6.2.1   Problem 65

Is it true that if $f'(c) > 0$ then there exists a neighbourhood of $c$ where $f$ is monotone increasing?

### 6.2.2   Problem 66

Answer the previous question under the additional assumption that $f'$ exists and is continuous in a neighbourhood of $c$.

### 6.2.3   Problem 67

(i) Are the conditions $\lim_{x \to +\infty} f(x) = +\infty$ and $\lim_{x \to +\infty} f'(x) = l < 0$ compatible?
(ii) Is it possible that a function $f$ satisfies the above conditions under the additional assumption that $f$ is continuous and differentiable in a neighbourhood of $+\infty$.

### 6.2.4 Problem 68

Is it possible that a function $f$ satisfies both conditions $\lim_{x \to +\infty} f(x) = +\infty$ and $\lim_{x \to +\infty} f'(x) = 0$?

### 6.2.5 Problem 69

Is it true that $\lim_{x \to +\infty} f'(x) = +\infty$ implies $\lim_{x \to +\infty} f(x) = +\infty$?

### 6.2.6 Problem 70

Give an example of a function defined on the interval $[-1, 1]$ which is bounded but without minimum and maximum points, and is continuous only at one point.

### 6.2.7 Problem 71

Give an example of a function defined on $\mathbb{R}$ such that every point of $\mathbb{R}$ is a local maximum point but not a global one.

### 6.2.8 Problem 72

Let $f$ be a real-valued function defined on the interval $(a, b)$. Assume that $P = (x_1, f(x_1))$ is a point of the graph of $f$. Let $t$ be the line tangent the graph of $f$ at $P$, see Fig. 6.1. Let $x_2 \in (a, b)$ with $x_2 > x_1$. Let $Q, R, S$ be three points with abscissa equal to $x_2$, where $R$ belongs to $t$, $S$ is above $t$ and $Q$ is below $t$. Assume that $f$ is twice differentiable and that $f''(x) > 0$ for all $x \in (a, b)$. Which one of the following three cases

(1) $Q$ belongs to the graph of $f$
(2) $R$ belongs to the graph of $f$
(3) $S$ belongs to the graph of $f$

is compatible with Lagrange's Theorem?

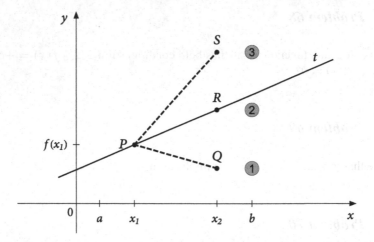

**Fig. 6.1** Problem 72

### 6.2.9   Problem 73

Let $f$ be a real-valued function defined on an open interval $I$ and let $c \in I$ be fixed. We say that $f$ is concave upward at $c$ if inequality (6.2) is satisfied for any $x_1, x_2 \in I$ sufficiently close to $c$, with $x_1 < c < x_2$.

(i)  Prove that if $f$ is concave upward at $c$ and differentiable at $c$ then the graph of $f$ is above the tangent line $y = f(c) + f'(c)(x - c)$ in a neighborhood of $c$.

(ii) Give an example of a function which is differentiable only at one point $x = c$ and such that its graph is above the tangent line $y = f(c) + f'(c)(x - c)$ on the right-hand side of $c$ and below that tangent line on the left-hand side of $c$.

### 6.2.10   Problem 74

Give an example of a function which is differentiable only at one point $x = c$ and is neither concave upward nor concave downward, neither on the right-hand side nor on the left-hand side of $c$.

### 6.2.11   Problem 75

Is $f''(x_0) = 0$ a sufficient condition for an inflection point at $x_0$?

## 6.2.12   Problem 76

If a curve $C : y = f(x)$ admits an inflection tangent line at a point $P$, does one
necessarily have $f''(x_P) = 0$?

# 6.3   Solutions

## 6.3.1   Solution 65

Not necessarily. Indeed, consider functions of the following type:

$f(x) = $ random $(x)$ with $2x - 1 \leq$ random $(x) \leq x^2$.

See Fig. 6.2.

By comparing any function $f$ with the functions $y = x^2$ and $y = 2x - 1$ (which
are differentiable at the point $x = 1$ with derivatives equal to 2) we get $f'(1) = 2$.
However, a function like $f$ is not necessarily monotone, as the following examples
show.

*Example 1*
$f(x) = $ random $(x)$, with $2x - 1 \leq$ random $(x) \leq x^2$, where $f$ is discontinuous at
any point $x \neq 1$. A function like this could be represented as in Fig. 6.2. Note that
the set of discontinuities of a monotone function can be only finite or countable.

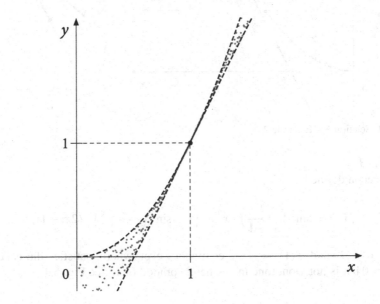

**Fig. 6.2** Solution 65 (Example 1)

*Example 2*

$$f(x) = \begin{cases} x^2, & \text{if } x \text{ is rational} \\ 2x - 1, & \text{if } x \text{ is irrational} \end{cases}$$

As in the general case, we have that $f'(1) = 2$. However, the function is not monotone increasing in any neighbourhood of 1 because, as one can see in the zoomed detail in Fig. 6.3, given a point $x = c$ with $c$ rational, it is possible to find infinitely many irrational numbers larger than $c$ where the function assume values less than $f(c)$.

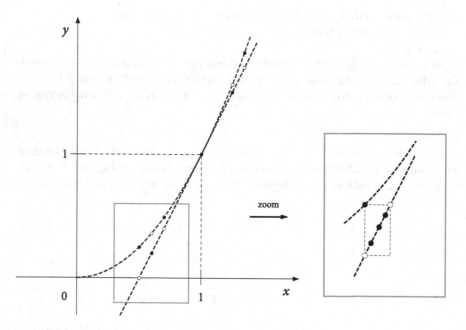

**Fig. 6.3** Solution 65 (Example 2)

*Example 3*
The function defined by

$$f(x) = \sin^2\left(\frac{9}{x-1}\right) \cdot x^2 + \left(1 - \sin^2\left(\frac{9}{x-1}\right)\right) \cdot (2x - 1),$$

for all $x \neq 1$, and $f(1) = 1$, is continuous everywhere, satisfies the condition $f'(1) > 0$ but is not monotone in any neighborhood of 1. See Fig. 6.4.

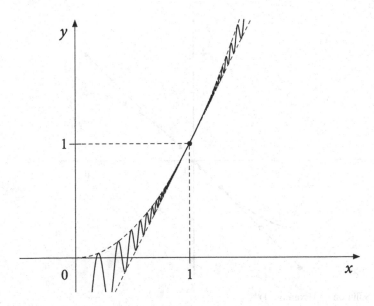

**Fig. 6.4** Solution 65 (Example 3)

*Example 4*
A different example in the spirit of Example 2 is provided by the function defined by

$$f(x) = \begin{cases} \sin x, & \text{if } -\frac{\pi}{2} \leq x \leq \frac{\pi}{2} \text{ with } x \text{ is rational} \\ x, & \text{if } -\frac{\pi}{2} \leq x \leq \frac{\pi}{2} \text{ with } x \text{ is irrational} \end{cases}$$

See Fig. 6.5.

## 6.3.2 Solution 66

True

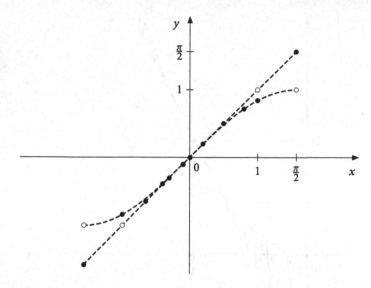

**Fig. 6.5** Solution 65 (Example 4)

### 6.3.3  Solution 67

(1)  Yes

  *Example*

$$f(x) = \begin{cases} -x + 1, & \text{if } 0 \le x < 1 \\ -x + 3, & \text{if } 1 \le x < 2 \\ -x + 5, & \text{if } 2 \le x < 3 \\ \cdots & \cdots \\ \cdots & \cdots \\ -x + 2n + 1, & \text{if } n \le x < n + 1 \end{cases}$$

  See Fig. 6.6.

(2)  No

### 6.3.4  Solution 68

Yes

*Example*  $y = \sqrt{x}, \quad y' = \frac{1}{2\sqrt{x}}$
  or
  $y = \log x, \quad y' = \frac{1}{x}$

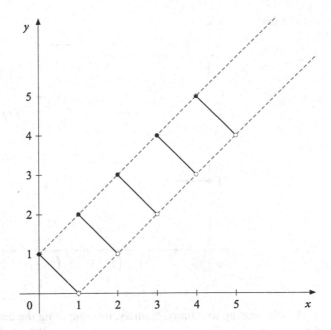

**Fig. 6.6**  Solution 67

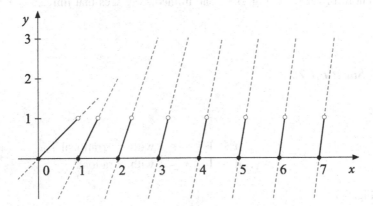

**Fig. 6.7**  Solution 69 (graph of the function)

### 6.3.5  *Solution 69*

Not necessarily.

*Example* $f(x) = nx - n^2 + n$ if $n - 1 \leq x < n - 1 + \frac{1}{n}$ where $n$ assumes the values of all non-zero natural numbers.

As one can easily see in Fig. 6.7, the graph of the function is given by an infinite number of segments (contained in the strip between by the lines $y = 0$ and $y = 1$),

**Fig. 6.8** Solution 69 (graph of the derivative)

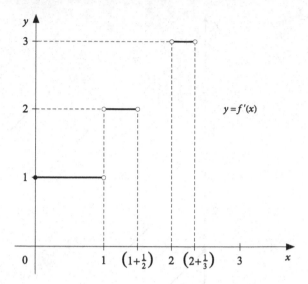

the slope of which is increasing to infinity. Namely, by computing the derivative of the function, that is $f'(x) = n$ if $n - 1 < x < n - 1 + \frac{1}{n}$, for all natural numbers $n$ different from zero (see Fig. 6.8), one immediately sees that $\lim_{x \to +\infty} f'(x) = +\infty$.

### 6.3.6   Solution 70

*Example*

$$f(x) = \begin{cases} x, & \text{if } -1 \leq x \leq 1 \text{ with } x \text{ irrational} \\ -\frac{x}{2}, & \text{if } -1 \leq x \leq 1 \text{ with } x \text{ rational} \end{cases}$$

See Fig. 6.9.

### 6.3.7   Solution 71

*Example*  $y = [x]$ = integer part of $x$
See Fig. 6.10.

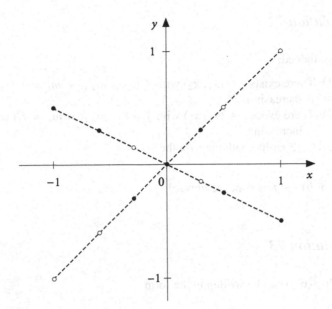

**Fig. 6.9** Solution 70

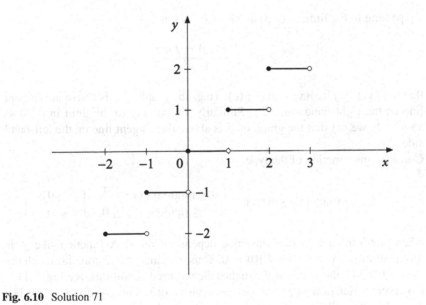

**Fig. 6.10** Solution 71

### 6.3.8   Solution 72

The case (3). Indeed:

>CASE (1): There exists $c \in (x_1, x_2)$ with $f'(c) = m_{PQ} < m_t = f'(x_1)$ ⨏
>because $f'$ is increasing.
>CASE (2): There exists $c \in (x_1, x_2)$ with $f'(c) = m_{PR} = m_t = f'(x_1)$ ⨏
>because $f'$ is increasing.
>CASE (3): Compatible with the hypothesis.

*Conclusion*
$f'' > 0$ on $(a, b) \to f$ is concave upward.

### 6.3.9   Solution 73

(i)  Inequality (6.2) can be written in the form

$$\frac{f(x_1) - f(c)}{x_1 - c} \leq \frac{f(x_2) - f(c)}{x_2 - c}. \tag{6.3}$$

By passing to the limit in (6.3) as $x_1 \to c^-$ we get

$$f'(c) \leq \frac{f(x_2) - f(c)}{x_2 - c},$$

that is $f(x_2) \geq f'(c)(x_2 - c) + f(c)$. Thus, the graph of $f$ is above the tangent line on the right-hand side of $c$. Similarly, by passing to the limit in (6.3) as $x_2 \to c^+$, we get that the graph of $f$ is above the tangent line on the left-hand side of $c$.

(ii)  Consider any function of the type

$$f(x) = \text{random}(x) \quad \text{where} \quad \begin{cases} 0 \leq \text{random}(x) \leq x^3, & \text{if } x \geq 0 \\ x^3 \leq \text{random}(x) \leq 0, & \text{if } x < 0 \end{cases}$$

where random$(x)$ is a random value depending on $x$. A function like $f$ is differentiable at $x = 0$ with $f'(0) = 0$. Thus the tangent line coincides with the $x$-axis. Clearly, the graph of $f$ satisfies the required conditions, see Fig. 6.11. However, a function like $f$ is not necessarily differentiable for $x \neq 0$. For example, the following function is not even continuous for $x \neq 0$ (see Fig. 6.12).

$$f(x) = \begin{cases} x^3, & \text{if } x \text{ is rational} \\ \frac{x^3}{2}, & \text{if } x \text{ is irrational} \end{cases}$$

**Fig. 6.11** Solution 73 (first example)

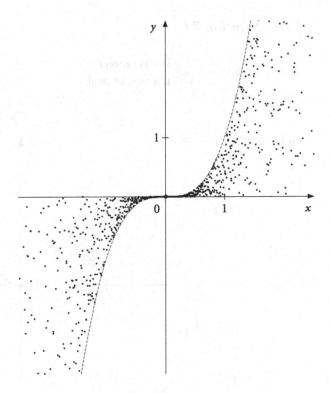

**Fig. 6.12** Solution 73 (second example)

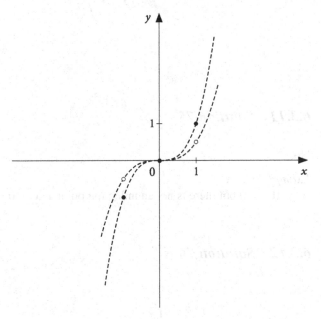

### 6.3.10   Solution 74

Example  $f(x) = \begin{cases} x^2, & \text{if } x \text{ is rational} \\ -x^2, & \text{if } x \text{ is irrational} \end{cases}$
   See Fig. 6.13.

**Fig. 6.13** Solution 74

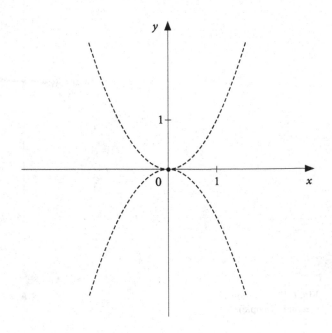

### 6.3.11   Solution 75

No.

Example  $y = x^4$
   $y''(0) = 0$ but there is not an inflection point at $x = 0$.

### 6.3.12   Solution 76

No.

Example  $C : y = \begin{cases} x^2 & \text{if } x \geq 0 \\ -x^2 & \text{if } x < 0 \end{cases}$

**Fig. 6.14** Solution 76

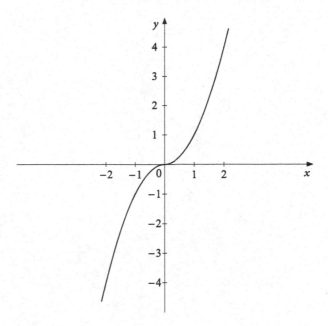

The $x$-axis is an inflection tangent line for the curve $C$ at the point 0 (see Fig. 6.14). However

$$+2 = y''_+(0) \neq y''_-(0) = -2 \quad \rightarrow \quad y''(0) \text{ does not exist}$$

By the Solutions 75 and 76 we deduce that $f''(x_0)$ is neither a necessary nor a sufficient condition for an inflection point.

# Chapter 7
# Graphs of functions

In this chapter we discuss the problem of plotting the graph of a given function, as well as the inverse problem of finding suitable formulas representing a given graph. It is a tradition in calculus to study the first problem, which is clearly important. However, nowadays the power of calculators has trivialized the first problem up to some extent, since any student can use a pocket size calculator to plot the graph of a function, defined by a given formula. Needless to say that the first problem is still of great pedagogical value. On the other hand, the ability to find a formula defining a function the graph of which is given, is also of great importance: this is useful not only in modelling a phenomenon but also at the very elementary level of preparing an exercise for students.

This chapter is organized as follows: first we recall the notion of asymptote of a function, second we describe a simple recipe for plotting the graph of an algebraic function based on elementary methods (such as the Euclidean division for polynomials, for example), finally we propose a number of examples which show how to find formulas starting from given graphs by using the same recipe.

## 7.1 Theoretical background

### 7.1.1 Asymptotes

**Definition 7.1 (Vertical asymptote)** Let $\mathcal{A}$ be a subset of $\mathbb{R}$ and $c \in \mathbb{R}$ be a limit point of $\mathcal{A}$. Let $f$ be a function from $\mathcal{A}$ to $\mathbb{R}$. We say that $f$ has a vertical asymptote at the point $c$ if

$$\lim_{x \to c} f(x) = \infty.$$

In this case, we also say that the line of equation $x = c$ is a vertical asymptote.

© The Author(s), under exclusive license to Springer Nature Switzerland AG 2022
P. Toni et al., *100+1 Problems in Advanced Calculus*, Problem Books
in Mathematics, https://doi.org/10.1007/978-3-030-91863-7_7

**Definition 7.2 (Horizontal or slant asymptotes at $+\infty$)** Let $\mathcal{A}$ be a subset of $\mathbb{R}$. Assume that $+\infty$ is a limit point of $\mathcal{A}$ (which means that $\mathcal{A}$ contains arbitrarily large positive numbers). Let $f$ be a function from $\mathcal{A}$ to $\mathbb{R}$. We say that $f$ has an asymptote at $+\infty$ if there exist $m, q \in \mathbb{R}$ such that

$$\lim_{x \to +\infty} (f(x) - (mx + q)) = 0\,.$$

In this case, we also say that the line of equation $y = mx + q$ is an horizontal asymptote if $m = 0$ and a slant asymptote if $m \neq 0$.

**Definition 7.3 (Horizontal or slant asymptotes at $-\infty$)** Let $\mathcal{A}$ be a subset of $\mathbb{R}$. Assume that $-\infty$ is a limit point of $\mathcal{A}$ (which means that $\mathcal{A}$ contains negative numbers, arbitrarily large in modulus). Let $f$ be a function from $\mathcal{A}$ to $\mathbb{R}$. We say that $f$ has an asymptote at $-\infty$ if there exist $m, q \in \mathbb{R}$ such that

$$\lim_{x \to -\infty} (f(x) - (mx + q)) = 0\,.$$

In this case, we also say that the line of equation $y = mx + q$ is a horizontal asymptote if $m = 0$ and a slant asymptote if $m \neq 0$.

It is simple to realize that the line of equation $y = mx + q$ is an asymptote at $+\infty$ if and only if

$$m = \lim_{x \to +\infty} \frac{f(x)}{x}, \quad \text{and} \quad q = \lim_{x \to +\infty} (f(x) - mx)\,.$$

The same statement holds true if we replace $+\infty$ by $-\infty$. This criterion is typically used in order to find possible asymptotes of functions. However, we shall see that a simpler method is available for a large class of algebraic functions.

The notion of asymptote can be extended in order to include other curves. For example, one may say that a function $f$ has a parabolic asymptote at $+\infty$ if there exist $a, b, c \in \mathbb{R}$ such that

$$\lim_{x \to +\infty} (f(x) - (ax^2 + bx + c) = 0\,,$$

in which case the parabola of equation $y = ax^2 + bx + c$ could be called asymptotic parabola at $+\infty$. In general, given a function $f$, another function $g$ could be considered as an asymptote of $f$ at $+\infty$ if

$$\lim_{x \to +\infty} (f(x) - g(x)) = 0\,. \tag{7.1}$$

This notion is clearly of interest only when the study of $g$ is simpler than the study of $f$.

*Remark 7.4* Assume that $g(x) \neq 0$. The reader must be aware of the fact that condition (7.1) is stronger than condition

$$\lim_{x \to +\infty} \frac{f(x)}{g(x)} = 1. \tag{7.2}$$

In fact (7.1) implies (7.2) but not viceversa. For example, the two functions $f(x) = x + \sin x$ and $g(x) = x$ satisfy condition (7.2) but not (7.1). Condition (7.2) is very important, and is much used in calculus: recall that if (7.2) holds then one also writes $f(x) \sim g(x)$ as $x \to +\infty$.

### 7.1.2 Hints on the degree, the asymptotic behaviour and the continuity of an algebraic curve

One of the main aims of this chapter is to show how it is possible, in some simple cases, to draw the graph of an algebraic function (rational or irrational) by using only the notions of degree, asymptote, continuity and the passing of the curve through a few points. By doing so, we shall try to point out the value of these important theoretical notions (degree, continuity) which do not involve much calculus but lead to noteworthy results. For example, we shall consider functions of the type:

$$y = \frac{A(x)}{B(x)}, \quad \text{or} \quad y = \sqrt{C(x)}$$

where $A, B, C$ are polynomials in $x$. These functions can be represented in the following form:

$$\begin{cases} B(x)y - A(x) = 0 \\ B(x) \neq 0 \end{cases} \quad \text{or} \quad \begin{cases} y^2 - C(x) = 0 \\ y \geq 0 \end{cases}$$

hence their graphs are given by one or more branches of algebraic curves of the type

$$P(x, y) = 0$$

where $P$ is a polynomial in $x, y$ of degree $n$ ($n$ is called the degree of the curve). For example, in the first case the graph of the function $y = \frac{A(x)}{B(x)}$ is the graph of the curve of equation $B(x)y - A(x) =$ with the exclusion of those point whose abscissas $x$ satisfy the equation $B(x) = 0$. In the second case the graph of the function $y = \sqrt{C(x)}$ is the graph of the curve of equation $y^2 - C(x) = 0$ with the exclusion of those points whose ordinates $y$ satisfy the condition $y < 0$.

We begin with a few general considerations.

**Degree**

The number of common points of an algebraic curve $P_n(x, y) = 0$ of degree $n > 1$ and an arbitrary line $r : ax + by + c = 0$ is at most $n$. Indeed, the system

$$\begin{cases} P_n(x, y) = 0 \\ ax + by + c = 0 \end{cases}$$

has at most $n$ solutions. In particular, if we are given a curve of degree $m + 1$ of the type

$$B_m(x)y = A_n(x), \quad \text{with} \quad m \geq n$$

where $A_n$ and $B_m$ are two polynomials in $x$ of degrees $n, m$ respectively, we can immediately conclude that it can intersects a horizontal line $r : y = c$ at most in $m$ points. Indeed, the system:

$$\begin{cases} B_m(x)y = A_n(x) \\ y = c \end{cases}$$

has at most $m$ solutions (the resolvent equation $B_m(x) - A_n(x) = 0$ has degree $\leq m$).

**Asymptotes**

In order to find the asymptotes or, more in general, the approximating curves of a given rational function of the type

$$y = \frac{A_n(x)}{B_m(x)}, \quad \text{in the case } n \geq m$$

it suffices to perform the Euclidean division of $A(x)$ by $B(x)$ and find the quotient $Q_{n-m}$ and the remainder $R_p$ (with $p < m$).

By recalling that

$$\text{dividend} = \text{quotient} \cdot \text{divisor} + \text{remainder}$$

hence

$$\frac{\text{dividend}}{\text{divisor}} = \text{quotient} + \frac{\text{remainder}}{\text{divisor}}$$

we can write the function in the following more useful form

$$y = Q_{n-m}(x) + \frac{R_p(x)}{B_m(x)}$$

We now observe that $\lim_{x \to \infty} \frac{R_p(x)}{B_m(x)} = 0$ (because $p < m$) hence the contribution of $\frac{R_p(x)}{B_m(x)}$ is negligible as $x \to \infty$. The curve can be approximated as follows

$$\boxed{y \simeq Q_{n-m}(x)}$$

as $x \to \infty$.

Moreover, it is easy to find the intersections between the curve and that asymptote: indeed, it suffices to solve the system

$$\begin{cases} y = Q_{n-m}(x) + \frac{R_p(x)}{B_m(x)} \\ y = Q_{n-m}(x) \end{cases}$$

which might have a high degree a priori, but always leads to an equation of smaller degree (namely, the degree of the remainder). Indeed, by means of a simple substitution, we get the resolvent equation

$$\boxed{R_p(x) = 0}$$

The corresponding solutions are the abscissas of the common points of the given function and the approximating curve $y = Q_{n-m}(x)$.

*Example 1*

$$y = \frac{x^3 + 1}{x^2 + 1} = x + \frac{1 - x}{x^2 + 1}$$

The slant asymptote is the line $y = x$ which intersects the curve at the point $x = 1$.

*Example 2*

$$y = \frac{x^3 + x^2 + 2}{x + 1} = x^2 + \frac{2}{x + 1}$$

The approximating curve is the parabola $y = x^2$ which does not intersect the graph of the given function.

Similarly, in the case of a rational curve of the type

$$y = \frac{A_n(x)}{B_m(x)}, \quad \text{with } n < m$$

we can observe that it can be viewed as

$$y = 0 + \frac{A_n(x)}{B_m(x)},$$

hence the horizontal asymptote is the line $y = 0$ while the possible intersections between the curve and the asymptote can be obtained by solving the equation

$$A_n(x) = 0$$

in the domain of the given function. (We note that this is nothing but a particular case of the previous one: indeed the Euclidean division between a polynomial $A_n$ of degree $n$ and a polynomial $B_m$ of degree $m$ with $n < m$, gives 0 as quotient and $A_n$ as remainder.)

### Continuity

We recall the following property of continuous functions.

Let $y = g(x)$ a continuous function whose graph divides the plane into two regions $\alpha$, $\beta$ and let $y = f(x)$ another continuous function. If the graph of $f$ passes through a point $A \in \alpha$ and a point $B \in \beta$ then it intersects the graph of $g$ at least at a point $C$ (see Fig. 7.1).

**Fig. 7.1** Continuity and border crossing

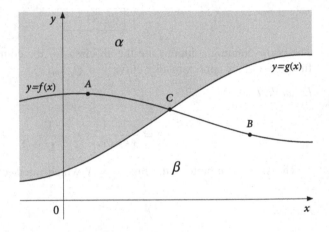

## 7.2   Problems

### 7.2.1   Problem 77

Is it possible that a function defined in the whole of $\mathbb{R}$ has a vertical asymptote?

### 7.2.2 Problem 78

Consider the function:

$$f(x) = \begin{cases} \frac{1}{x} & \text{if } 2n < x \leq 2n+1 \\ \frac{1}{x} + 1 & \text{if } 2n+1 < x \leq 2n+2 \end{cases}$$

where $n$ assumes all values of $\mathbb{N}$ (see Fig. 7.2).

Does $f$ prove that a function defined on the set of positive real numbers may have two horizontal asymptotes?

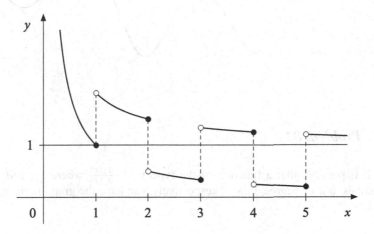

**Fig. 7.2** Problem 78

### 7.2.3 Problem 79

Give an example of a function $f$ defined on the set of non-negative real numbers, differentiable everywhere possibly with the exception of a countable set of points, such that $f'(x) = 0$ for all $x$ where $f'(x)$ exists, and such that the line $y = x$ is a slant asymptote of it.

### 7.2.4 Problem 80

Is it possible for a nowhere differentiable function to have an asymptote?

### 7.2.5    Problem 81

Say if it is possible that a polynomial function of degree 3 of the type $y = ax^3 + bx^2 + cx + d$ can have a graph of the type represented in Fig. 7.3.

**Fig. 7.3**  Problem 81

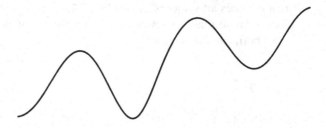

### 7.2.6    Problem 82

Say if it is possible that a function of the type $y = \frac{A_3(x)}{B_2(x)}$ where $A_3$ and $B_2$ are polynomials in $x$ of degree 3 and 2 respectively, can have the graph represented in Fig. 7.4.

**Fig. 7.4**  Problem 82

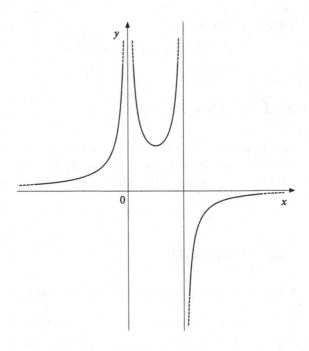

### 7.2.7 Problem 83

Say if it is possible that a function of the type $y = \frac{A_2(x)}{B_2(x)}$ where $A_2$ and $B_2$ are polynomials in $x$ of degree 2, can have the graph represented in Fig. 7.5.

**Fig. 7.5** Problem 83

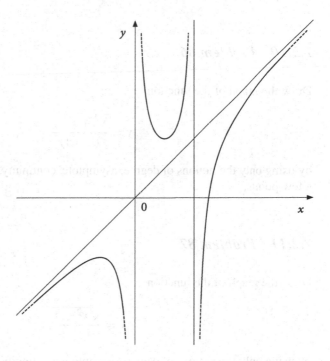

### 7.2.8 Problem 84

Draw the graph of the function

$$y = \frac{x^2 + x - 3}{x^2 - 1}$$

by using only the notions of degree, asymptote, continuity, and the passage through a few points.

### 7.2.9 Problem 85

Draw the graph of the function

$$y = \frac{x^3 - 2}{2x}$$

by using only the notions of degree, asymptote, continuity, and the passage through a few points.

### 7.2.10   Problem 86

Draw the graph of the function

$$y = \frac{x^2 + x + 2}{(x + 1)^2}$$

by using only the notions of degree, asymptote, continuity, and the passage through a few points.

### 7.2.11   Problem 87

Draw the graph of the function

$$y = \frac{\sqrt{x^2 + 1}}{x - 2}$$

by using only the notions of degree, asymptote, continuity, and the passage through a few points.

### 7.2.12   Problem 88

Draw the graph of the function

$$y = \arcsin \frac{2x}{x^2 + 1} - 2 \arctan x$$

### 7.2.13   Problem 89

Draw the graph of the function

$$y = \arccos \frac{x^2 - 1}{x^2 + 1} + 2 \arctan x$$

### 7.2.14 Problem 90

Draw the graph of the function

$$y = \arcsin \frac{x}{\sqrt{x^2 + 1}} - \arctan x$$

## 7.3 Solutions

### 7.3.1 Solution 77

Yes

*Example 1*
See Fig. 7.6.

**Fig. 7.6** Solution 77
(Example 1)

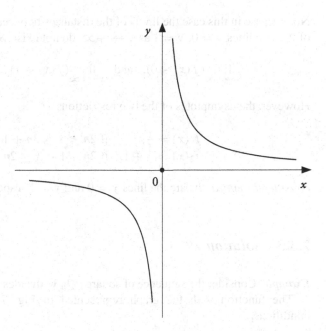

*Example 2*
See Fig. 7.7.

**Fig. 7.7** Solution 77
(Example 2)

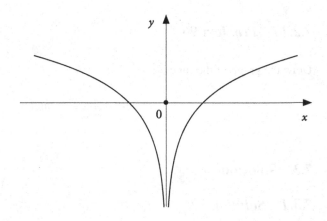

### 7.3.2   Solution 78

No, because in this case the limits of the distances between the graph of $f$ and each of the two lines $y = 0$, $y = 1$ as $x \to +\infty$, do not exist. Namely, we have that

$$\lim_{x \to +\infty} (f(x) - 0), \quad \text{and} \quad \lim_{x \to +\infty} (f(x) - 1) \ \text{do not exist}.$$

However, the asymptotes of the two restrictions

$$f_1(x) = \tfrac{1}{x}, \qquad \text{if } 2n < x \le 2n + 1$$
$$f_2(x) = \tfrac{1}{x} + 1, \quad \text{if } 2n + 1 < x \le 2n + 2,$$

*considered separately*, are the lines $y = 0$ and $y = 1$ respectively.

### 7.3.3   Solution 79

*Example* Consider the sequence of squares $Q_n$ with sides $\tfrac{1}{n}$ represented in Fig. 7.8;
    The function with the graph represented in Fig. 7.9 satisfies the required conditions.

**Fig. 7.8** Solution 79 (the sequence of squares)

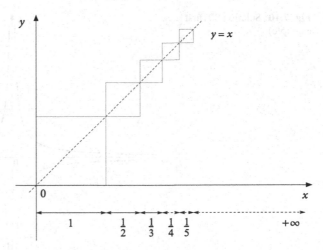

**Fig. 7.9** Solution 79 (the graph of the function)

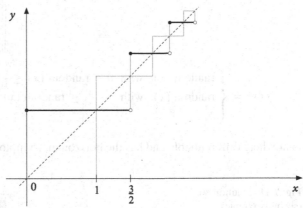

### 7.3.4 Solution 80

Yes.

*Examples* The function defined by

$$f(x) = \begin{cases} \frac{1}{x}, & \text{if } x \in \mathbb{Q}, \\ 0, & \text{if } x = 0, \\ -\frac{1}{x}, & \text{if } x \in \mathbb{R} \setminus \mathbb{Q} \end{cases}$$

is nowhere differentiable and has two asymptotes: the vertical asymptote with equation $x = 0$ and the horizontal asymptote with equation $y = 0$ (see Fig. 7.10).

Also the function defined by

**Fig. 7.10** Solution 80 (first example)

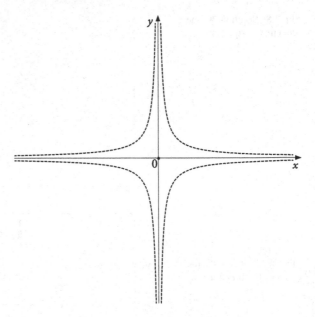

$$f(x) = \begin{cases} \text{random } (x), \text{ with } 0 < \text{random } (x) \le \frac{1}{|x|}, & \text{if } x \in \mathbb{Q} \setminus \{0\}, \\ \text{random } (x), \text{ with } -\frac{1}{|x|} \le \text{random } (x) < 0, & \text{if } x \in \mathbb{R} \setminus \mathbb{Q}, \\ 1, & \text{if } x = 0 \end{cases}$$

is nowhere differentiable and has the horizontal asymptote $y = 0$ (see Fig. 7.11).

**Fig. 7.11** Solution 80 (second example)

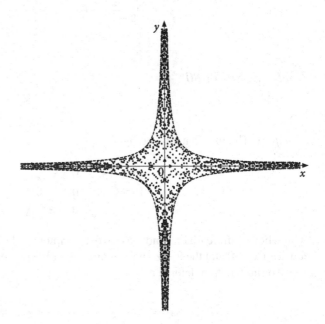

### 7.3.5 Solution 81

No. Indeed, there would exist a line as in Fig. 7.12 that would intersect the curve at 5 points which is impossible because the curve is of degree 3.

**Fig. 7.12** Solution 81

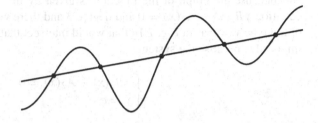

### 7.3.6 Solution 82

No, because the graph of the function is given by three branches of the curve of equation $yB_2(x) - A_3(x) = 0$ and there would exist a line as in Fig. 7.13 that would intersect the curve at 4 points which is impossible because the curve is of degree 3.

**Fig. 7.13** Solution 82

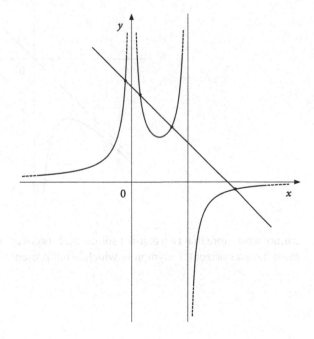

Moreover, it is evident that the function must have a slant asymptote which is not present in the picture.

### 7.3.7  Solution 83

No, because the graph of the function is given by three branches of the curve of equation $y B_2(x) - A_2(x) = 0$ and degree 3 and there would exist a horizontal line of the type $y = c$ as in Fig. 7.14 that would intersect that curve at 3 points, which is impossible because the system

$$\begin{cases} y B_2(x) - A_2(x) = 0 \\ y = c \end{cases}$$

**Fig. 7.14**  Solution 83

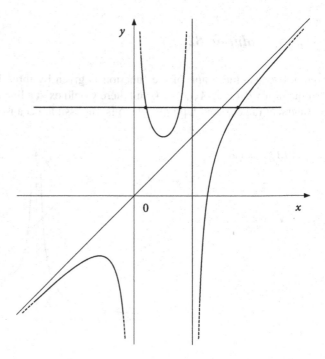

cannot have more that two distinct solutions. Moreover, it is evident that the function must have a horizontal asymptote which is not present in the picture.

## 7.3.8 Solution 84

**DEGREE**

The function can be written in the equivalent form:

$$\begin{cases} y(x^2 - 1) = x^2 + x - 3 \\ x^2 - 1 \neq 0 \end{cases}$$

that is

$$\begin{cases} yx^2 - x^2 - x - y + 3 = 0 \\ x \neq \pm 1 \end{cases}$$

We note that there are no points with abscissa $x = \pm 1$ which satisfy the equation $yx^2 - x^2 - x - y + 3 = 0$, hence the condition $x \neq \pm 1$ is superfluous. Thus the graph of the function is the graph of the curve of equation

$$yx^2 - x^2 - x - y + 3 = 0$$

This curve has degree 3 (see the table) hence the graph of the function can intersect any line in at most three points.

$$\boxed{\begin{array}{ccccc} yx^2 & -x^2 & -x & -y & +3 = 0 \\ \downarrow & \downarrow & \downarrow & \downarrow & \downarrow \\ \text{MAX} \{3, & 2, & 1, & 1, & 0\} = 3 \text{ (degree of the curve)} \end{array}}$$

**ASYMPTOTES**

By dividing the polynomial at the numerator by the polynomial at the denominator and computing the corresponding quotient and remainder, the function can be written in the form:

$$y = \frac{x^2 + x - 3}{x^2 - 1} = 1 + \frac{x - 2}{(x + 1)(x - 1)}$$

Thus:

- The horizontal asymptote is the line $y = 1$ which intersects the curve only at the point $x = 2$.
- The vertical asymptotes are the lines $x = \pm 1$.

**Fig. 7.15** Solution 84 Step A

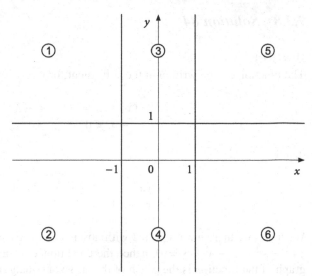

## CONTINUITY

The function is continuous on its domain of definition, that is, for $x \neq \pm 1$, because it is the ratio of two continuous functions.

## GRAPH

In order to draw the graph we proceed in the following way.

(A)  We draw the asymptotes: the plane is divided into the 6 regions represented in Fig. 7.15.

(B)  We begin with choosing any point through which the curve passes, for example $P = (-2, -\frac{1}{3})$, in Fig. 7.16. Point $P$ belongs to the region (2), hence the curve must stay in the region (2) for $x < -1$ because it crosses the horizontal asymptote only at the point $A = (2, 1)$. Moreover, the fact the line $x = -1$ is a vertical asymptote and that the curve must stay in the region (2) for $x < -1$, allows us to conclude that $\lim_{x \to -1^-} f(x) = -\infty$ (without computing it).

(C)  As far as the strip between the two vertical asymptotes is concerned, considering that the curve does not cross the horizontal asymptote in the strip, we have to choose between two possible cases (see Fig. 7.17).
We can discard Case II because if Case II would occur then there would exist a line as in Fig. 7.17 that would intersect the curve at 4 points, which is impossible because the curve has degree 3. Thus we are left with Case I only.

(D)  As far as the half plane $x > 1$ is concerned, we choose for simplicity a point through which the curve passes: for example, $Q = (3, \frac{9}{8})$. By considering the passage through $Q$, the vertical asymptote $x = 1$, and the horizontal asymptote $y = 1$ (which is crossed by the curve only at the point $A = (2, 1)$), we see that we have to choose between two possible cases (see Fig. 7.18).

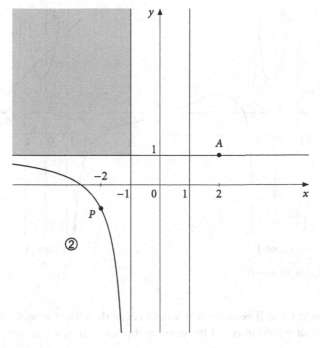

**Fig. 7.16** Solution 84 Step B

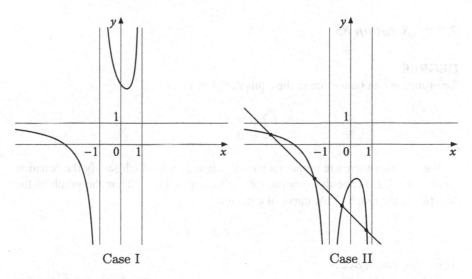

Case I                    Case II

**Fig. 7.17** Solution 84 Step C

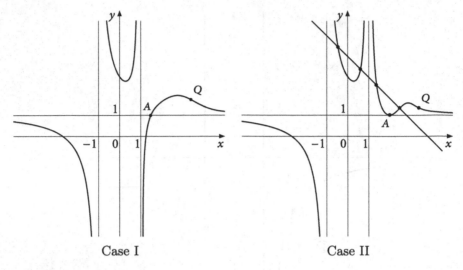

**Fig. 7.18** Solution 84 Step D

We can discard Case II because if it would occur then there would exist a line as in Fig. 7.18 that would intersect the curve at 4 points. Thus we are left with Case I only.

### 7.3.9   Solution 85

**DEGREE**

The function can be written in the equivalent form:

$$\begin{cases} 2xy = x^3 - 2 \\ x \neq 0 \end{cases}$$

We note that there are no points with abscissa $x = 0$ which satisfy the equation $2xy = x^3 - 2$, hence the condition $x \neq 0$ is superfluous. Thus the graph of the function is the graph of the curve of equation

$$x^3 - 2xy - 2 = 0$$

This curve has degree 3.

$$\begin{array}{ccc} x^3 & -2xy & -2 = 0 \\ \downarrow & \downarrow & \downarrow \\ \text{MAX } \{3, & 2, & 0, \} = 3 \text{ (degree of the curve)} \end{array}$$

## ASYMPTOTES

By dividing the polynomial at the numerator by the polynomial at the denominator and computing the corresponding quotient and remainder, the function can be written in the form:

$$y = \frac{x^3 - 2}{2x} = \frac{x^2}{2} - \frac{1}{x}$$

Thus:

- The approximating curve is the parabola $y = \frac{x^2}{2}$ which does not intersect the curve.
- The vertical asymptote is the line $x = 0$.

## CONTINUITY

The function is continuous on its domain of definition, that is, for $x \neq 0$, because it is the ratio of two continuous functions.

## GRAPH

In order to draw the graph we proceed in the following way.

(A) We draw the vertical asymptote $x = 0$ and the approximating curve $y = \frac{x^2}{2}$: the plane is divided into the 4 regions represented in Fig. 7.19.

(B) We begin with choosing any point through which the curve passes, for example $P = (-1, \frac{3}{2})$, in Fig. 7.20. Point $P$ belongs to the region (1), hence the curve must stay in the region (1) for $x < 0$ because it does not crosses the parabola.

**Fig. 7.19** Solution 85 Step A

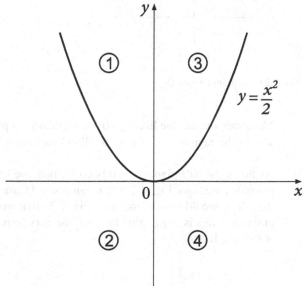

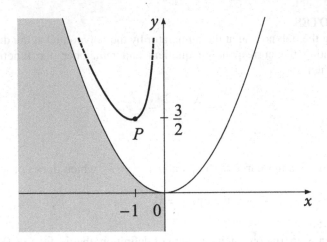

**Fig. 7.20**  Solution 85 Step B

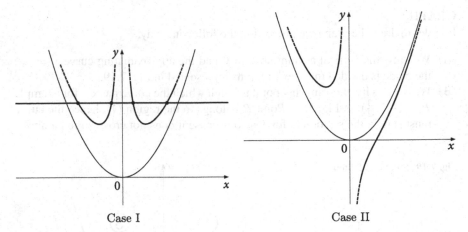

Case I                                              Case II

**Fig. 7.21**  Solution 85 Step C

Moreover, the fact the line $x = 0$ is a vertical asymptote and that the curve must stay in the region (1) for $x < 0$, allows us to conclude that $\lim_{x \to 0^-} f(x) = +\infty$.

(C) As far as the half plane $x > 0$ is concerned, we have to choose between two possible cases (see Fig. 7.21). We can discard Case I because if it would occur then there would exist a line as in Fig. 7.21 that would intersect the curve at 4 points and this is impossible because the curve has degree 3. Thus we are left with Case II only.

## 7.3.10   Solution 86

**DEGREE**
The function can be written in the equivalent form:

$$(x + 1)^2 y = x^2 + x + 2$$

that is

$$x^2 y - x^2 + 2xy + y - x - 2 = 0$$

Thus its graph is the graph of a curve of degree 3.

$$
\begin{array}{cccccc}
x^2 y & -x^2 & +2xy & +y & -x & -2 = 0 \\
\downarrow & \downarrow & \downarrow & \downarrow & \downarrow & \downarrow \\
\text{MAX } \{3, & 2, & 2, & 1, & 1, & 0\} = 3 \text{ (degree of the curve)}
\end{array}
$$

Thus the graph of the function can intersects any line at most 3 points. In this case, we can say even more: since the equation of the curve is in the form $(x + 1)^2 y = x^2 + x + 2$, then the graph of the function can intersect any *horizontal* line in at most 2 points (see page 150 with $m = n = 2$). Namely, the possible intersections of the curve with a line $y = c$ are obtained by solving the system

$$
\begin{cases}
(x + 1)^2 y = x^2 + x + 2 \\
y = c
\end{cases}
$$

the resolvent equation of which is

$$(x + 1)^2 c = x^2 + x + 2$$

which has degree 2 hence it cannot have more than 2 solutions.

**ASYMPTOTES**
By dividing the polynomial at the numerator by the polynomial at the denominator and computing the corresponding quotient and remainder, the function can be written in the form:

$$y = \frac{x^2 + x + 2}{(x + 1)^2} = 1 + \frac{1 - x}{(x + 1)^2}$$

Thus:

- The horizontal asymptote is the line $y = 1$ which intersects the curve only at the point $x = 1$.
- The vertical asymptote is the line $x = -1$.

**Fig. 7.22**  Solution 86 Step A

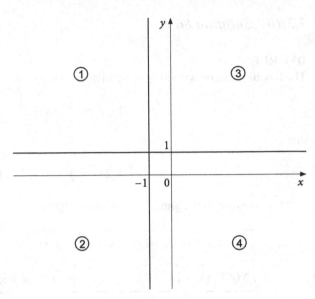

**CONTINUITY**

The function is continuous on its domain of definition, that is, for $x \neq -1$.

**GRAPH**

In order to draw the graph we proceed in the following way.

(A)  We draw the asymptotes: the plane is divided into the 4 regions represented in Fig. 7.22.

(B)  We begin with choosing any point through which the curve passes, for example $P = (-2, 4)$. Point $P$ belongs to the region (1), hence the curve must stay in the region (1) for $x < -1$ because it crosses the horizontal asymptote only at the point $A = (1, 1)$. Moreover, since $\lim_{x \to -1} f(x) = \infty$ and the curve cannot stay in the region (2), we immediately have that $\lim_{x \to -1^-} f(x) = +\infty$. Thus, the graph for $x < -1$ is as in Fig. 7.23.

(C)  As far as the half plane $x > -1$ is concerned, we have to choose between the 4 possible cases represented in Fig. 7.24, which are all compatible with the passage through $A = (1, 1)$, with the vertical asymptote $x = 0$ and with the horizontal asymptote $y = 1$. We can discard Cases II, III, IV because if they would occur then there would exist a horizontal line as in Fig. 7.24 that would intersect the curve at more than 2 points. Thus we are left with Case I only.

**Fig. 7.23** Solution 86 Step B

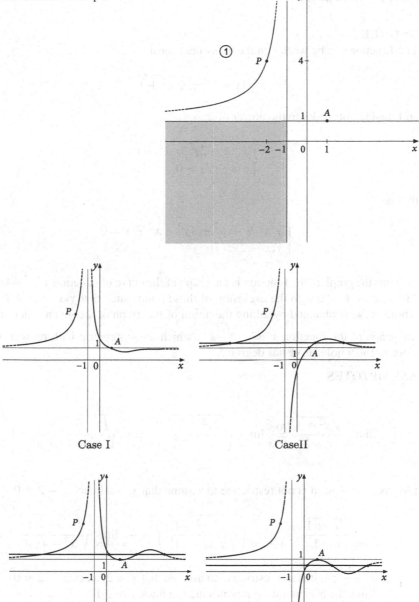

**Fig. 7.24** Solution 86 Step C

### 7.3.11   Solution 87

**DEGREE**

The function can be written in the equivalent form:

$$y(x - 2) = \sqrt{x^2 + 1}$$

and raising both sides to the power of 2:

$$\begin{cases} y^2(x - 2)^2 = x^2 + 1 \\ y(x - 2) > 0 \end{cases}$$

that is

$$\begin{cases} y^2 x^2 - 4xy^2 + 4y^2 - x^2 - 1 = 0 \\ y(x - 2) > 0 \end{cases}$$

Thus the graph of the function is the graph of the curve of equation $y^2 x^2 - 4xy^2 + 4y^2 - x^2 - 1 = 0$ with the exclusion of those points such that $y(x - 2) < 0$. The whole curve is obtained by taking the union of the graph of the given function and the graph of the function $y = -\frac{\sqrt{x^2+1}}{x-2}$, which are symmetric with respect to the $x$-axis. The whole curve has degree 4.

**ASYMPTOTES**

$$\lim_{x \to +\infty} \frac{\sqrt{x^2 + 1}}{x - 2} \overset{(A)}{=} \lim_{x \to +\infty} \sqrt{\frac{x^2 + 1}{(x - 2)^2}} = \lim_{x \to +\infty} \sqrt{\frac{x^2 + 1}{x^2 - 4x + 4}} = 1$$

(A) as $x \to +\infty$, it is not restrictive to assume that $x > 2$ hence $x - 2 > 0$

$$\lim_{x \to -\infty} \frac{\sqrt{x^2 + 1}}{x - 2} \overset{(B)}{=} \lim_{x \to +\infty} \left(-\sqrt{\frac{x^2 + 1}{(x - 2)^2}}\right) = -\lim_{x \to -\infty} \sqrt{\frac{x^2 + 1}{x^2 - 4x + 4}} = -1$$

(B) as $x \to -\infty$, it is not restrictive to assume that $x < 2$ hence $x - 2 < 0$
  Thus the horizontal asymptotes are the lines $y = \pm 1$.

$$\lim_{x \to 2} \frac{\sqrt{x^2 + 1}}{x - 2} = \infty$$

The vertical asymptote is the line $x = 2$.

## INTERSECTIONS OF THE FUNCTION WITH THE HORIZONTAL ASYMPTOTES

(1) *Intersections with the asymptote $y = 1$*

$$\frac{\sqrt{x^2+1}}{x-2} = 1$$
$$\sqrt{x^2+1} = x - 2 \quad \text{(hence } x - 2 > 0)$$
$$x^2 + 1 = x^2 - 4x + 4$$
$$x = \tfrac{3}{4} \text{ not compatible with the condition } x - 2 > 0$$

(2) *Intersections with the asymptote $y = -1$*

$$\frac{\sqrt{x^2+1}}{x-2} = -1$$
$$\sqrt{x^2+1} = 2 - x \quad \text{(hence } 2 - x > 0)$$
$$x^2 + 1 = x^2 - 4x + 4$$
$$x = \tfrac{3}{4} \text{ compatible with the condition } 2 - x > 0$$

Thus, the graph of the function intersects the asymptote $y = -1$ at the point $x = \tfrac{3}{4}$.

## GRAPH

In order to draw the graph we proceed in the following way.

(A) We draw the asymptotes: the plane is divided into the 6 regions represented in Fig. 7.25.

(B) We begin with choosing any point through which the curve passes, for example $P = (3, \sqrt{10})$. Point $P$ belongs to the region (1), hence the curve must stay in the region (1) for $x > 2$ because it does not cross the horizontal asymptote $y = 1$. Thus, the graph for $x > 2$ is as in Fig. 7.26.

(C) As far as the half plane $x < 2$ is concerned, keeping in mind the passage through $A = (\tfrac{3}{4}, -1)$, the asymptotes $y = -1$ and $x = 2$, we have to choose between the 2 possible cases represented in Fig. 7.27.

We can discard Case II because if it would occur then there would exist a line as in Fig. 7.28 that would intersect the whole curve (obtained by considering also the graph of the function $y = -\frac{\sqrt{x^2+1}}{x-2}$) at 6 points which is against the fact that the curve has degree 4. Thus we are left with Case I only.

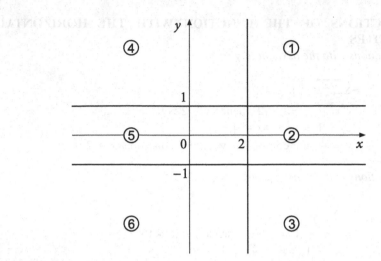

**Fig. 7.25** Solution 87 Step A

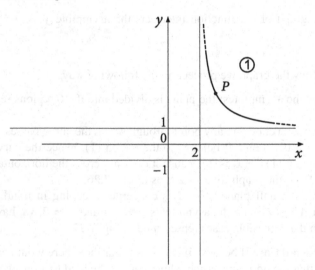

**Fig. 7.26** Solution 87 Step B

Thus, the function $y = \frac{\sqrt{x^2+1}}{x-2}$ has the graph represented in Fig. 7.29.

The graph of the function $y = -\frac{\sqrt{x^2+1}}{x-2}$ is symmetric (see Fig. 7.30).

The graph of the whole curve $y^2x^2 - 4xy^2 + 4y^2 - x^2 - 1 = 0$ is the union of the two graphs above, represented in Fig. 7.31.

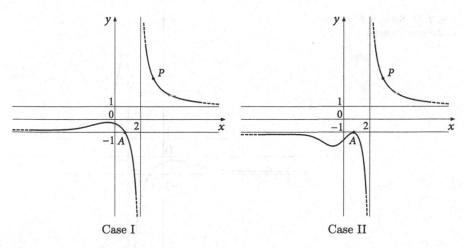

Case I                                      Case II

**Fig. 7.27** Solution 87 Step C

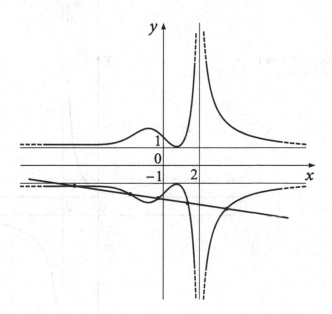

**Fig. 7.28** Solution 87 (discarting case II)

**Fig. 7.29** Solution 87 (graph
of the given function)

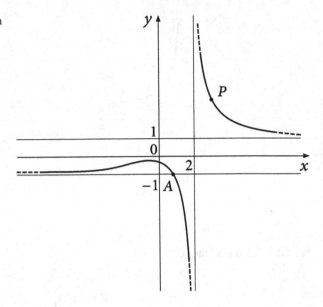

**Fig. 7.30** Solution 87 (graph
of the opposite function)

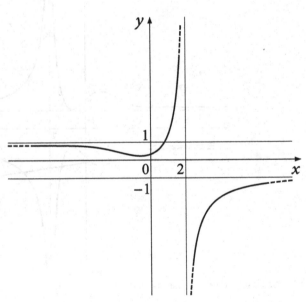

**Fig. 7.31** Solution 87 (graph of the whole curve)

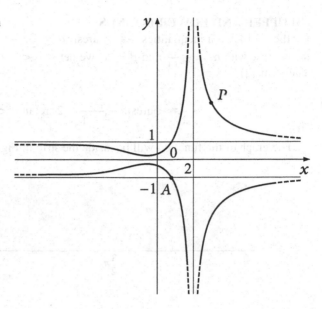

### 7.3.12 Solution 88

#### (A) DOMAIN

The only condition that the variable $x$ has to satisfy is $-1 \leq \frac{2x}{x^2+1} \leq 1$. Let us solve this system:

$$-1 \leq \frac{2x}{x^2+1} \leq 1$$

$\Big\uparrow$ multiplying by $x^2 + 1$ (Theorem $T_3$) $\Big\downarrow$

$$-x^2 - 1 \leq 2x \leq x^2 + 1$$

$\Big\uparrow$ summing $-2x$ (Theorem $T_1$) $\Big\downarrow$

$$-x^2 - 2x - 1 \leq 0 \leq x^2 - 2x + 1$$

$\Big\uparrow$ factoring $\Big\downarrow$

$$-(x+1)^2 \leq 0 \leq (x-1)^2 \quad \textit{always fullfilled}$$

Thus the domain is $\mathbb{R}$.

## (B) UPPER AND LOWER BOUNDS

By the well-known inequalities $-\frac{\pi}{2} \leq \arcsin\alpha \leq \frac{\pi}{2}$, $-\frac{\pi}{2} < \arctan\beta < \frac{\pi}{2}$ applied to our case with $\alpha = \frac{2x}{x^2+1}$ and $\beta = x$, we get $-\pi < -2\arctan x < \pi$ and (using Theorem $T_4$)

$$-\frac{3}{2}\pi < \arcsin\frac{2x}{x^2+1} - 2\arctan x < \frac{3}{2}\pi$$

The graph of the function will be inside the strip in Fig. 7.32.

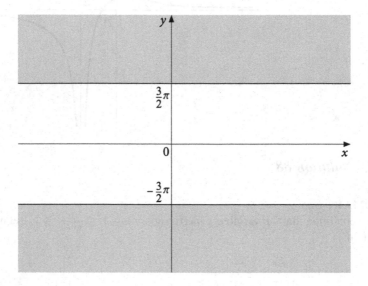

**Fig. 7.32**  Solution 88 Step B

## (C) SYMMETRIES

The function is odd because:

$$f(-x) = \arcsin\frac{-2x}{(-x)^2+1} - 2\arctan(-x)$$

$$= -\arcsin\frac{2x}{x^2+1} + 2\arctan x = -f(x)$$

## (D) ASYMPTOTES

$$\lim_{x\to\pm\infty}\left(\arcsin\frac{2x}{x^2+1} - 2\arctan x\right)$$

$$= \lim_{x\to\pm\infty}\arcsin\frac{2x}{x^2+1} - 2\lim_{x\to\pm\infty}\arctan x = 0 - 2\left(\pm\frac{\pi}{2}\right) = \mp\pi$$

Thus the horizontal asymptotes are the lines $y = \mp \pi$, while there are no vertical asymptotes because the function is bounded.

## (E) CONTINUITY
The function is continuous because it is the sum and the composition of continuous functions.

## (F) DERIVATIVE

$$y'(x) = \frac{1}{\sqrt{1 - \left(\frac{2x}{x^2+1}\right)^2}} \cdot \frac{2(x^2+1) - 4x^2}{(x^2+1)^2} - \frac{2}{x^2+1}$$

$$= \sqrt{\left(\frac{x^2+1}{x^2-1}\right)^2} \cdot \frac{2(1-x^2)}{(x^2+1)^2} - \frac{2}{x^2+1}$$

$$= \frac{x^2+1}{|x^2-1|} \cdot \frac{2(1-x^2)}{(x^2+1)^2} - \frac{2}{x^2+1}$$

$$= \begin{cases} \frac{-4}{x^2+1} & \text{if } x < -1 \text{ or } x > 1 \\ 0 & \text{if } -1 < x < 1 \end{cases}$$

Thus:

- If $x < -1$ or $x > 1$ then $y'(x) = \frac{-4}{x^2+1} < 0$ and the function is decreasing.
- If $-1 < x < 1$ then $y'(x) = 0$ and the function is constant.

## (G) CORNER POINTS
The function is not differentiable at $x = \pm 1$ where it has two corner points. Indeed:

$$\begin{cases} \lim_{x \to 1^+} f'(x) = \lim_{x \to 1^+} \frac{-4}{x^2+1} = -2 \\ \lim_{x \to 1^-} f'(x) = \lim_{x \to 1^-} 0 = 0 \end{cases}$$

and, as one can guess from the symmetry of the function,

$$\begin{cases} \lim_{x \to -1^+} f'(x) = \lim_{x \to -1^+} 0 = 0 \\ \lim_{x \to -1^-} f'(x) = \lim_{x \to -1^-} \frac{-4}{x^2+1} = -2 \end{cases}$$

Thus, since $y(1) = 0$ and $y(-1) = 0$, the corner points are: $A = (1, 0)$, and $B = (-1, 0)$.

## (H) INTERSECTIONS WITH THE y-AXIS
Since the function is odd, then $f(x) = -f(-x)$ for all $x$. Thus, for $x = 0$ we get $f(0) = -f(-0) = -f(0)$ hence $2f(0) = 0$, that is $f(0) = 0$.

## (I) GRAPH
The graph of the function is represented in Fig. 7.33.

**Fig. 7.33** Solution 88 (graph of the given function)

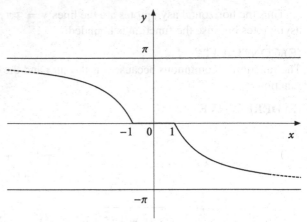

### 7.3.13   Solution 89

#### (A) DOMAIN

The only condition that the variable $x$ has to satisfy is $-1 \leq \frac{x^2-1}{x^2+1} \leq 1$. Let us solve this system:

$$-1 \leq \frac{x^2 - 1}{x^2 + 1} \leq 1$$

multiplying by $x^2 + 1$ (Theorem $T_3$)

$$-(x^2 + 1) \leq x^2 - 1 \leq x^2 + 1$$

summing 1 (Theorem $T_1$)

$$-x^2 \leq x^2 \leq x^2 + 2 \quad \textit{always fullfilled}$$

Thus the domain is $\mathbb{R}$.

## (B) UPPER AND LOWER BOUNDS

By the well-known inequalities $0 \leq \arccos \alpha \leq \pi$, $-\frac{\pi}{2} \leq \arctan \beta \leq \frac{\pi}{2}$ applied to our case with $\alpha = \frac{x^2-1}{x^2+1}$ and $\beta = x$, we get $-\pi < 2 \arctan x < \pi$ and (using Theorem T$_4$)

$$-\pi < \arccos \frac{x^2-1}{x^2+1} + 2 \arctan x < 2\pi$$

The graph of the function will be inside the strip in Fig. 7.34.

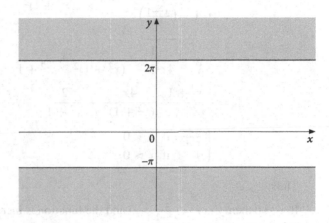

**Fig. 7.34** Solution 89 Step B

## (C) SYMMETRIES

The function is neither odd nor even because:

$$f(-x) = \arccos \frac{(-x)^2-1}{(-x)^2+1} - 2 \arctan(-x)$$

$$= \arccos \frac{x^2-1}{x^2+1} - 2 \arctan x \neq \begin{cases} f(x) \\ -f(x) \end{cases}$$

## (D) ASYMPTOTES

$$\lim_{x \to \pm\infty} \left( \arccos \frac{x^2-1}{x^2+1} + 2 \arctan x \right)$$

$$= \lim_{x \to \pm\infty} \arccos \frac{x^2-1}{x^2+1} + 2 \lim_{x \to \pm\infty} \arctan x = 0 + 2 \left( \pm\frac{\pi}{2} \right) = \pm\pi$$

Thus the horizontal asymptotes are the lines $y = \pm\pi$, while there are no vertical asymptotes because the function is bounded.

## (E) CONTINUITY
The function is continuous because it is the sum and the composition of continuous functions.

## (F) DERIVATIVE

$$y'(x) = \frac{-1}{\sqrt{1 - \left(\frac{x^2-1}{x^2+1}\right)^2}} \cdot \frac{2x(x^2+1) - (x^2-1)2x}{(x^2+1)^2} + \frac{2}{x^2+1}$$

$$= -\sqrt{\frac{(x^2+1)^2}{4x^2}} \cdot \frac{4x}{(x^2+1)^2} + \frac{2}{x^2+1}$$

$$= -\frac{x^2+1}{2|x|} \cdot \frac{4x}{(x^2+1)^2} + \frac{2}{x^2+1}$$

$$= \begin{cases} \frac{4}{x^2+1} & \text{if } x < 0 \\ 0 & \text{if } x > 0 \end{cases}$$

Thus:

- If $x < 0$ then $y'(x) = \frac{4}{x^2+1} > 0$ and the function is increasing.
- If $x > 0$ then $y'(x) = 0$ and the function is constant.

## (G) CORNER POINTS
The function is not differentiable at $x = 0$ where it has a corner point. Indeed:

$$\begin{cases} \lim_{x \to 0^-} f'(x) = \lim_{x \to 0^-} \frac{4}{x^2+1} = 4 \\ \lim_{x \to 0^+} f'(x) = \lim_{x \to 0^+} 0 = 0 \end{cases}$$

Thus, since $y(0) = \pi$, the corner point is $A = (0, \pi)$.

## (I) GRAPH
The graph of the function is represented in Fig. 7.35.

*Remark* The functions in Problems 88 and 89 are particularly interesting because their graphs have corner points even though no absolute values appear in the formulas defining the functions, as often happen when corner points appear. Moreover, many students overlook a problem of sign and forget the use of an absolute value in the computation of the derivative.

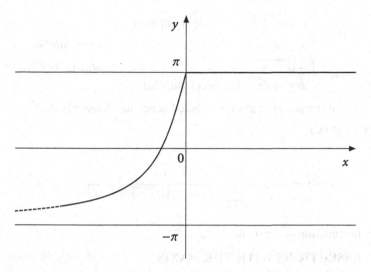

**Fig. 7.35** Solution 89 (graph of the given function)

## 7.3.14 Solution 90

### (A) DOMAIN

The only condition that the variable $x$ has to satisfy is $-1 \leq \dfrac{x}{\sqrt{x^2+1}} \leq 1$. Let us solve this system:

$$-1 \leq \frac{x}{\sqrt{x^2 + 1}} \leq 1$$

$\Big\uparrow$ multiplying by $\sqrt{x^2 + 1}$ (Theorem T$_3$)

$$-\sqrt{x^2 + 1} \leq x \leq \sqrt{x^2 + 1}$$

$$\begin{cases} -\sqrt{x^2 + 1} \leq x \\ x \leq \sqrt{x^2 + 1} \end{cases}$$

$-$ If $x \geq 0$ then $\begin{cases} -\sqrt{x^2+1} \leq x & \text{\textit{always fulfilled}} \\ x \leq \sqrt{x^2+1} \xleftrightarrow{T_8} x^2 \leq x^2+1 & \text{\textit{always fulfilled}} \end{cases}$

$-$ If $x < 0$ then $\begin{cases} -\sqrt{x^2+1} \leq x \xleftrightarrow{T_9} x^2+1 \geq x^2 & \text{\textit{always fulfilled}} \\ x \leq \sqrt{x^2+1} & \text{\textit{always fulfilled}} \end{cases}$

Thus the initial inequality is always satisfied hence the domain is $\mathbb{R}$.

**(B) DERIVATIVE**

$$y'(x) = \frac{1}{\sqrt{1 - \frac{x^2}{x^2+1}}} \cdot \frac{1}{(x^2+1)\sqrt{x^2+1}} - \frac{1}{x^2+1} = 0$$

Thus the function is constant.

**(C) INTERSECTIONS WITH THE $y$-AXIS**

$y(0) = \arcsin 0 - \arctan 0 = 0$

Thus, $y(x) = 0$ for all $x$.

*Remark* This exercise can be considered as a simple way to prove the trigonometric identity

$$\arcsin \frac{x}{\sqrt{x^2+1}} = \arctan x$$

## 7.3.15   Step-by-step construction of a function with a given graph

We present a number of examples of curves defined step-by-step using from time to time the notions that we have already discussed for algebraic curves. In general, the sequence of steps is the following:

(A)  Choose an asymptote $\begin{cases} \text{horizontal} \\ \text{slant} \\ \text{parabolic} \begin{cases} \text{with vertical axis} \\ \text{with horizontal axis} \end{cases} \end{cases}$

(B)  Choose possible vertical asymptotes

(C)  Specify the lateral behaviour of the function at those asymptotes

(D)  Choose the possible intersection points of the curve with the asymptote chosen in step (A)

(E)  Other special conditions $\begin{cases} \text{passage through some given points} \\ \text{sign of the gap between the curve} \\ \text{and the asymptote chosen in (A)} \end{cases}$

*Remark* Given a function $y = f(x)$ and an asymptote $y = g(x)$, the gap between the curve and the asymptote is the new function $y = f(x) - g(x)$ which represents the signed measure of the oriented segment $QP$ in Fig. 7.36.

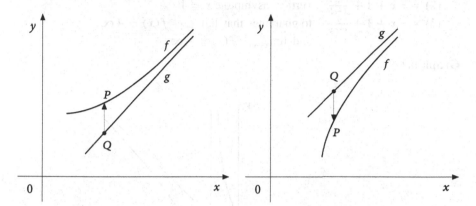

**Fig. 7.36** Gap between curve and asymptote

**Graph n.1**

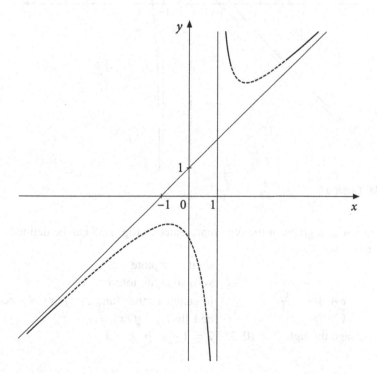

**Fig. 7.37** Graph n.1

A function with a graph of the type represented in Fig. 7.37 can be defined step by step as follows.

(1) $y = x + 1 + \ldots \ldots$ slant asymptote

(2) $y = x + 1 + \frac{\cdots}{x-1}$    vertical asymptote $x = 1$

(3) $y = x + 1 + \frac{1}{x-1}$    to guarantee that $\lim_{x \to 1^+} f(x) = +\infty$

and $\lim_{x \to 1^-} f(x) = -\infty$

**Graph n.2**

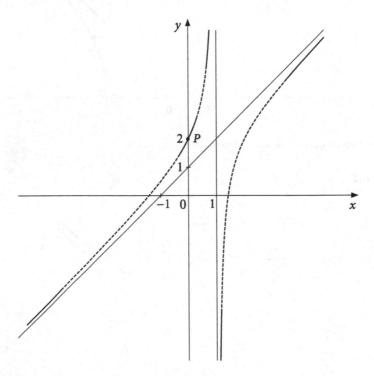

**Fig. 7.38**  Graph n.2

A function with a graph of the type represented in Fig. 7.38 can be defined step by step as follows.

(1) $y = x + 1 + \ldots \ldots$                 slant asymptote

(2) $y = x + 1 + \frac{\cdots}{x-1}$                 vertical asymptote $x = 1$

(3) $y = x + 1 + \frac{-k}{x-1}$                 to guarantee that $\lim_{x \to 1^+} f(x) = -\infty$

with $k > 0$                 and $\lim_{x \to 1^-} f(x) = +\infty$

(4) passage through $P = (0, 2) : 2 = 1 + k \to k = 1$

**Graph n.3**

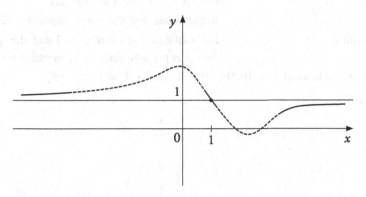

**Fig. 7.39** Graph n.3

A function with a graph of the type represented in Fig. 7.39 can be defined step by step as follows.

(1) $y = 1 + \ldots\ldots$ horizontal asymptote
(2) $y = 1 + \frac{\cdots}{x^2+1}$ there are no vertical asymptotes
(3) $y = 1 + \frac{1-x}{x^2+1}$ the curve intersects the horizontal asymptote at $x = 1$
and the gap $\frac{1-x}{x^2+1}$ is negative for $x > 1$, positive for $x < 1$

**Graph n.4**

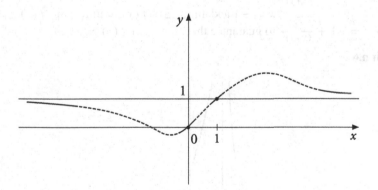

**Fig. 7.40** Graph n.4

A function with a graph of the type represented in Fig. 7.40 can be defined step by step as follows.

(1) $y = 1 + \ldots\ldots$          horizontal asymptote

(2) $y = 1 + \frac{\ldots}{x^2+1}$        there are no vertical asymptotes

(3) $y = 1 + \frac{k(x-1)}{x^2+1}$      to guarantee that  the curve intersects the

     with $k > 0$              horizontal asymptote at $x = 1$ and  the  gap

                              $\frac{k(x-1)}{x^2+1}$ is positive for $x > 1$,  negative for $x < 1$

(4) passage through $0 = (0, 0) : \ 0 = 1 - k \to \ k = 1$

**Graph n.5**

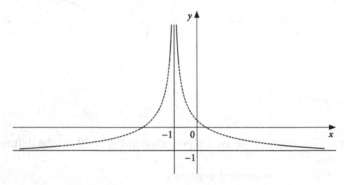

**Fig. 7.41**  Graph n.5

A function with a graph of the type represented in Fig. 7.41 can be defined step by
step as follows.

(1) $y = -1 + \ldots\ldots$ horizontal asymptote

(2) $y = -1 + \frac{\ldots}{(x+1)^2}$ to guarantee that  the vertical asymptote is

                $x = -1$ and $\lim_{x \to -1+} f(x) = \lim_{x \to -1-} f(x)$

(4) $y = -1 + \frac{1}{(x+1)^2}$ to guarantee that  $\lim_{x \to -1} f(x) = +\infty$

**Graph n.6**

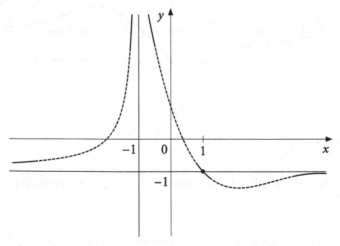

**Fig. 7.42**  Graph n.6

A function with a graph of the type represented in Fig. 7.42 can be defined step by step as follows.

  (1) $y = -1 + \ldots\ldots$ horizontal asymptote
  (2) $y = -1 + \frac{\ldots}{(x+1)^2}$ to guarantee that the vertical asymptote is

  $x = -1$ and $\lim_{x \to -1^+} f(x) = \lim_{x \to -1^-} f(x)$

  (3) $y = -1 + \frac{1-x}{(x+1)^2}$ to guarantee that the curve intersects the

  asymptote $y = -1$ at $x = 1$ and $\lim_{x \to -1} f(x) = +\infty$

**Graph n.7**

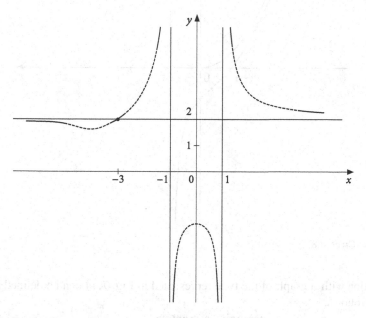

**Fig. 7.43**  Graph n.7

A function with a graph of the type represented in Fig. 7.43 can be defined step by step as follows.

  (1) $y = 2 + \ldots\ldots$    horizontal asymptote
  (2) $y = 2 + \frac{\ldots}{(x-1)(x+1)}$ vertical asymptotes $x \pm 1$
  (3) $y = 2 + \frac{x+3}{(x-1)(x+1)}$ the curve intersects the horizontal asymptote at

  $x = -3$ and the gap $\frac{x+3}{(x-1)(x+1)}$

  is negative for $x < -3$ and $-1 < x < 1$ and
  positive for $-3 < x < -1$ and $x > 1$

**Graph n.8**

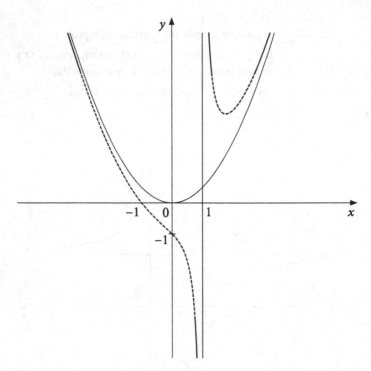

**Fig. 7.44**  Graph n.8

A function with a graph of the type represented in Fig. 7.44 can be defined step by step as follows.

(1) $y = \frac{1}{2}x^2 + \ldots\ldots$ parabolic asymptote

(2) $y = \frac{1}{2}x^2 + \frac{\ldots}{x-1}$  vertical asymptote $x = 1$

(3) $y = \frac{1}{2}x^2 + \frac{1}{x-1}$  the gap $\frac{1}{x-1}$ is positive for $x > 1$, hence the curve is above the parabola. In the same way, the curve is below the parabola for $x < 1$

A few bundles to which the curve belongs:

$\mathcal{F}_1 : y = \frac{1}{2}x^2 + \frac{a}{x-1}$

$\mathcal{F}_2 : y = ax^2 + \frac{1}{x-1}$

$\mathcal{F}_3 : y = \frac{1}{2}x^2 + \frac{1}{ax-1}$

$\mathcal{F}_4 : y = \frac{1}{2}x^2 + \frac{1}{x-a}$

**Graph n.9**

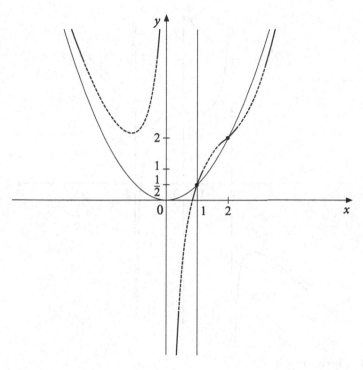

**Fig. 7.45** Graph n.9

A function with a graph of the type represented in Fig. 7.45 can be defined step by step as follows.

(1) $y = \frac{1}{2}x^2 + \ldots\ldots$      parabolic asymptote

(2) $y = \frac{1}{2}x^2 + \frac{\ldots}{x^m}$      to guarantee that the vertical asymptote
    with $m$ odd               is $x = 0$ and $\lim_{x \to 0^+} f(x) = -\lim_{x \to 0^-} f(x)$
    and $m \geq 3$

(3) $y = \frac{1}{2}x^2 + \frac{-(x-1)(x-2)}{x^m}$ to guarantee that the curve intersects
                              the parabolic asymptote at $x = 1$ and $x = 2$
                              and $\lim_{x \to 0^+} f(x) = -\infty$

**Graph n.10**

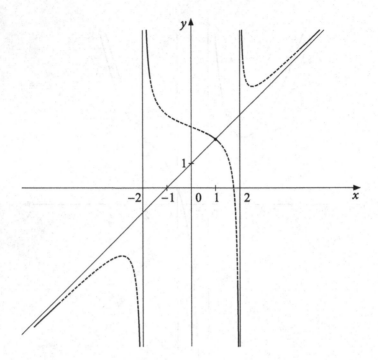

**Fig. 7.46** Graph n.10

A function with a graph of the type represented in Fig. 7.46 can be defined step by step as follows.

(1) $y = x + 1 + \ldots\ldots$     slant asymptote

(2) $y = x + 1 + \frac{\ldots}{(x-2)(x+2)}$   vertical asymptotes $x \pm 2$

(3) $y = x + 1 + \frac{x-1}{(x-2)(x+2)}$ to guarantee that the curve intersects
$y = x + 1$ at $x = 1$ and the gap $\frac{x-1}{(x-2)(x+2)}$
is positive if $-2 < x < 1$ and $x > 2$,
and negative if $x < -2$ and $1 < x < 2$

**Graph n.11**

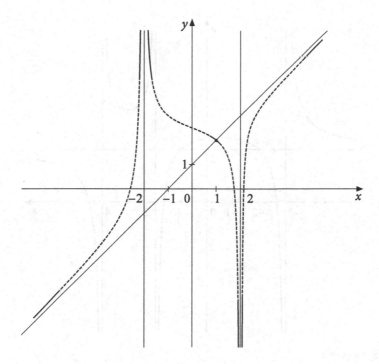

**Fig. 7.47** Graph n.11

A function with a graph of the type represented in Fig. 7.47 can be defined step by step as follows.

    (1) $y = x + 1 + \ldots \ldots$         slant asymptote

    (2) $y = x + 1 + \frac{\ldots}{(x-2)^m(x+2)^n}$ to guarantee that the vertical asymptotes

        with $m, n > 0$ both even    are $x = \pm 2$, $\lim_{x \to 2^+} f(x) = \lim_{x \to 2^-} f(x)$,

                             and $\lim_{x \to -2^+} f(x) = \lim_{x \to -2^-} f(x)$

    (3) $y = x + 1 + \frac{1-x}{(x-2)^m(x+2)^n}$ to guarantee that the curve intersects the

                             asymptote $y = x + 1$ at $x = 1$ and

                             $\lim_{x \to 2} f(x) = -\infty$, $\lim_{x \to -2} f(x) = +\infty$

**Graph n.12**

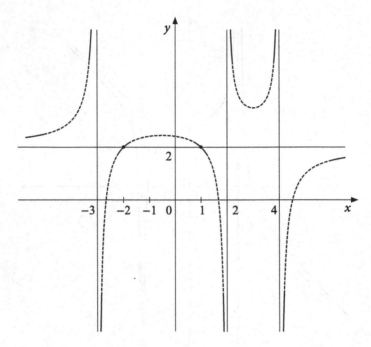

**Fig. 7.48**  Graph n.12

A function with a graph of the type represented in Fig. 7.48 can be defined step by step as follows.

(1) $y = 2 + \ldots \ldots$            horizontal asymptote

(2) $y = 2 + \frac{\ldots}{(x+3)(x-2)(x-4)}$   vertical asymptotes $x = 2$, $x = -3$, $x = 4$

(3) $y = 2 + \frac{-(x+2)(x-1)}{(x+3)(x-2)(x-4)}$   to guarantee that the curve intersects the asymptote $y = 2$ at $x = 1$ and $x = -2$ and the gap $\frac{-(x+2)(x-1)}{(x+3)(x-2)(x-4)}$ is positive if $x < -3$, $-2 < x < 1$, $2 < x < 4$ and negative if $-3 < x < 2$, $1 < x < 2$, $x > 4$

**Graph n.13**

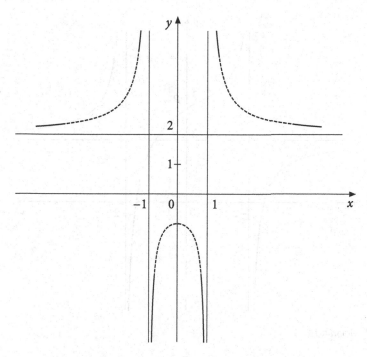

**Fig. 7.49** Graph n.13

A function with a graph of the type represented in Fig. 7.49 can be defined step by step as follows.

    (1) $y = 2 + \ldots\ldots$        horizontal asymptote

    (2) $y = 2 + \dfrac{\ldots}{(x-1)(x+1)}$    vertical asymptotes $x = \pm 1$

    (3) $y = 2 + \dfrac{1}{(x-1)(x+1)}$    to guarantee that the curve does not

                                    intersects the asymptote $y = 2$ and the

                                    gap $\dfrac{1}{(x-1)(x+1)}$ is positive if $x < -1$,

                                    $x > 1$, and negative if $-1 < x < 1$

**Graph n.14**

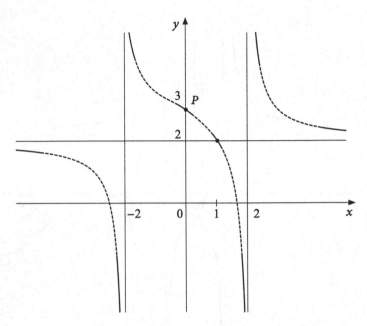

**Fig. 7.50** Graph n.14

A function with a graph of the type represented in Fig. 7.50 can be defined step by step as follows.

(1) $y = 2 + \ldots\ldots$          horizontal asymptote

(2) $y = 2 + \frac{\cdots}{(x+2)(x-2)}$  vertical asymptotes $x = \pm 2$

(3) $y = 2 + \frac{k(x-1)}{(x+2)(x-2)}$  to guarantee that the curve intersects

    with $k > 0$          the horizontal asymptote at $x = 1$

                    and the gap $\frac{k(x-1)}{(x+2)(x-2)}$ is positive if $-2 < x < 1$,

                    $x > 2$, and negative if $x < -2$, $1 < x < 2$

(4) passage through    $P = (0, 3) : 3 = 2 + \frac{k}{4} \;\rightarrow\; k = 4$

**Graph n.15**

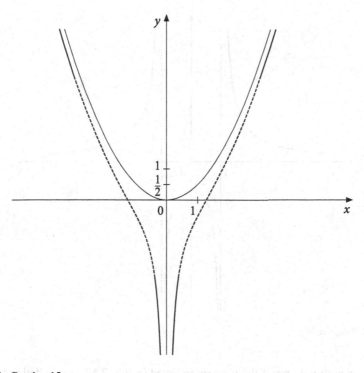

**Fig. 7.51** Graph n.15

A function with a graph of the type represented in Fig. 7.51 can be defined step by step as follows.

(1) $y = \frac{1}{2}x^2 + \ldots\ldots$ parabolic asymptote

(2) $y = \frac{1}{2}x^2 + \frac{\ldots}{x^n}$  to guarantee that the vertical asymptote

with $n > 0$ even   is $x = 0$ and $\lim_{x \to 0^+} f(x) = \lim_{x \to 0^-} f(x)$

(3) $y = \frac{1}{2}x^2 + \frac{-1}{x^n}$  to guarantee that $\lim_{x \to 0} f(x) = -\infty$

**Graph n.16**

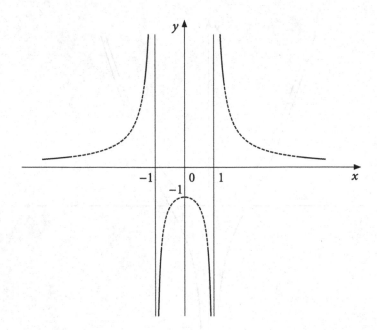

**Fig. 7.52** Graph n.16

A function with a graph of the type represented in Fig. 7.52 can be defined step by step as follows.

(1) $y = 0 + \ldots\ldots$           asymptote $x$-axis

(2) $y = 0 + \frac{\ldots}{x^2-1}$         vertical  asymptotes $x = \pm 1$

(3) $y = 0 + \frac{1}{x^2-1} = \frac{1}{x^2-1}$ to  guarantee  the passage  through $(0, -1)$

                                  and no intersections with $x$-axis

A few bundles to which the curve belongs:

$$\mathcal{F}_1 : \ y = \frac{k}{x^2-1}$$
$$\mathcal{F}_2 : \ y = \frac{1}{x^2-k}$$

Draw those bundles depending on $k \in \mathbb{R}$.

**Graph n.17**

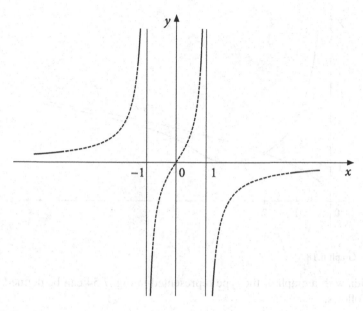

**Fig. 7.53** Graph n.17

A function with a graph of the type represented in Fig. 7.53 can be defined step by step as follows.

(1) $y = 0 + \ldots\ldots$  asymptote $x$-axis

(2) $y = 0 \ldots \frac{\cdots}{x^2-1}$  vertical asymptotes $x = \pm 1$

(3) $y = 0 \ldots \frac{x}{x^2-1}$  curve passing through the origin, odd function, graph symmetric with respect to the origin

(4) $y = 0 - \frac{x}{x^2-1}$  $\lim_{x \to +\infty} y = 0^-$, the curve converges asymptotically to the $x$-axis from below, $y < 0$ if $x > 1$

$$y = \frac{x}{1-x^2}$$

A few bundles to which the curve belongs:

$$y = \frac{kx}{1-x^2}; \quad y = \frac{x+k}{1-x^2}; \quad y = \frac{x}{k-x^2}; \quad y = \frac{kx}{k-x^2}$$

**Graph n.18**

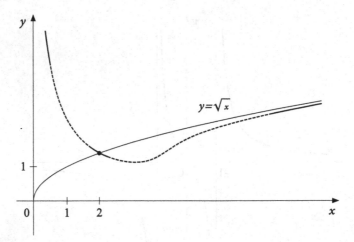

**Fig. 7.54** Graph n.18

A function with a graph of the type represented in Fig. 7.54 can be defined step by step as follows.

(1) $y = \sqrt{x} + \ldots\ldots$ approximating curve

(2) $y = \sqrt{x} \ldots \frac{\cdots}{x^2}$   vertical  asymptote $x = 0$

(3) $y = \sqrt{x} + \frac{2-x}{x^2}$   to guarantee that  the curve  intersects  the graph of the function  $y = \sqrt{x}$ at $x = 2$ and the gap $\frac{2-x}{x^2}$ is positive if  $x < 2$, negative if $x > 2$

**Graph n.19**

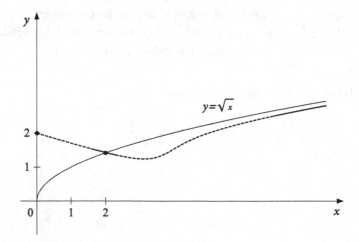

**Fig. 7.55** Graph n.19

A function with a graph of the type represented in Fig. 7.55 can be defined step by step as follows.

(1) $y = \sqrt{x} + \ldots\ldots$         approximating curve

(2) $y = \sqrt{x} + \frac{\ldots}{(x+1)^2}$      the function has no vertical asymptotes

(3) $y = \sqrt{x} + \frac{k(2-x)}{(x+1)^2}$      to guarantee that the curve intersects the
     with $k > 0$           graph of the function $y = \sqrt{x}$ at $x = 2$
                   and the gap $\frac{k(2-x)}{(x+1)^2}$ is positive if $x < 2$,

(4) passage through $P = (0, 2) : 2 = 2k \rightarrow k = 1$

**Graph n.20**

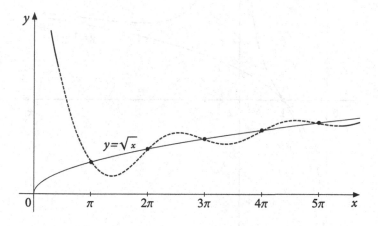

**Fig. 7.56** Graph n.20

A function with a graph of the type represented in Fig. 7.56 can be defined step by step as follows.

(1) $y = \sqrt{x} + \ldots\ldots$ approximating curve

(2) $y = \sqrt{x} + \frac{\sin x}{\ldots\ldots}$   to guarantee that the curve intersects the
                 graph of the function $y = \sqrt{x}$ at the
                 points $x = n\pi$ with $n \in \mathbb{N}_0$

(3) $y = \sqrt{x} + \frac{\sin x}{x^2}$   to guarantee that $\lim_{x \to +\infty}(f(x) - \sqrt{x}) = 0$,
                 $\lim_{x \to 0^+} f(x) = +\infty$ and that the gap $\frac{\sin x}{x^2}$ is
                 positive if $2n\pi < x < (2n + 1)\pi$ and negative
                 if $(2n + 1)\pi < x < (2n + 2)\pi$

**Graph n.21**

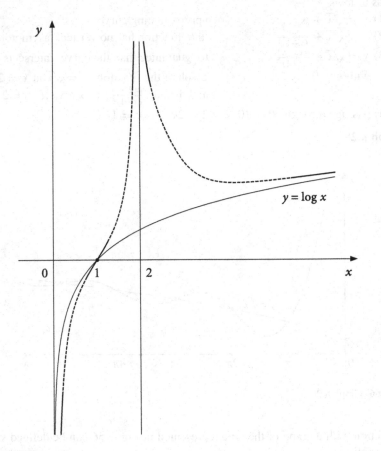

**Fig. 7.57**  Graph n.21

A function with a graph of the type represented in Fig. 7.57 can be defined step by step as follows.

(1) $y = \log x + \ldots\ldots$ approximating curve

(2) $y = \log x + \frac{\ldots\ldots}{(x-2)^4}$ vertical asymptote $x = 2$

(3) $y = \log x + \frac{x(x-1)}{(x-2)^4}$ to guarantee that the curve intersects the graph of the function $y = \log x$ at $x = 1$ and that the gap $\frac{x(x-1)}{(x-2)^4}$ is positive if $x > 1$ and negative if $x < 1$, and $\lim_{x\to 0^+} \frac{x(x-1)}{(x-2)^4} = 0^-$

# Chapter 8
# Integrals

## 8.1 Theoretical background

### 8.1.1 Areas and Riemann Integrals

Given a function $f$ from a bounded interval $[a, b]$ to $\mathbb{R}$, the Riemann integral $\int_a^b f(x)dx$ represents the signed area of the planar surface between the graph of $f$ and the $x$-axis, with the understanding that the negative values of $f$ contribute with a negative sign to value of the area. At a very elementary level, this could be taken as a definition, at least for sufficiently regular functions, say continuous. As soon as one considers functions with one or more points of discontinuity, the need of a more precise definition arises. In the case of bounded functions, there are at least three equivalent ways of defining the Riemann integral $\int_a^b f(x)dx$: one based on Cauchy sums, another based on Riemann sums and a third one based on the notion of area in the sense of Peano-Jordan. For the convenience of the reader, here we recall the latter.

Given a bounded interval $[a, b]$ in $\mathbb{R}$, we denote by $P$ a finite subset of $[a, b]$ including $a, b$. The set $P$ will be represented as $P = \{x_0, x_1, \ldots, x_n\}$ with

$$a = x_0 < x_1 < \cdots < x_n = b.$$

We also denote by $\mathcal{P}$ the collection of all subsets $P$ of $[a, b]$ as above.

Given a bounded real-valued function $f$ defined on a bounded interval $[a, b]$ and $P \in \mathcal{P}$, with $P = \{x_0, x_1, \ldots, x_n\}$, we set

$$s(P) = \sum_{k=0}^{n-1} \left( (x_{k+1} - x_k) \inf_{x \in [x_k, x_{k+1}]} f(x) \right) \qquad (8.1)$$

© The Author(s), under exclusive license to Springer Nature Switzerland AG 2022
P. Toni et al., *100+1 Problems in Advanced Calculus*, Problem Books
in Mathematics, https://doi.org/10.1007/978-3-030-91863-7_8

and

$$S(P) = \sum_{k=0}^{n-1} \left( (x_{k+1} - x_k) \sup_{x \in [x_k, x_{k+1}]} f(x) \right).$$ (8.2)

It is clear that the sum in (8.1) represents the area of the union of rectangles with width equal to $x_{k+1} - x_k$ and height equal to $\inf_{x \in [x_k, x_{k+1}]} f(x)$, for $k = 0, \ldots, n-1$. The union of those rectangles defines a so-called pluri-rectangle. In particular, for positive functions, those pluri-rectangles are "inscribed" in the planar region between the graph of $f$ and the $x$-axis. In the same way, the sum in (8.2) is associated with "circumscribed" pluri-rectangles.

We denote by $H$, $K$ the sets of all areas of all "inscribed" and "circumscribed" pluri-rectangles as above. Namely,

$$H = \{s(P) : P \in \mathcal{P}\}, \text{ and } K = \{S(P) : P \in \mathcal{P}\}.$$ (8.3)

Then we can give the following definition.

**Definition 8.1** Let $f$ be a bounded real-valued function defined on a bounded interval $[a, b]$. We say that $f$ is Riemann integrable if

$$\sup H = \inf K.$$ (8.4)

In that case, the Riemann integral of $f$ over $[a, b]$ is defined by

$$\int_a^b f(x)dx = \sup H = \inf K.$$

One basic criterion for establishing whether a function is Riemann integrable and for computing the integral is the following.

**Definition 8.2** Let $f$ be a bounded real-valued function defined on a bounded interval $[a, b]$. Then $f$ is Riemann integrable if and only if there exists a sequence $P_n \in \mathcal{P}$, for $n \in \mathbb{N}$, such that

$$\lim_{n \to \infty} s(P_n) = \lim_{n \to \infty} S(P_n),$$

in which case

$$\int_a^b f(x)dx = \lim_{n \to \infty} s(P_n) = \lim_{n \to \infty} S(P_n).$$

Note that taking the supremum of all possible sums $s(P)$ in (8.4) corresponds to the classical method of exhaustion for the computation of areas, a method which was known since ancient times. In fact, the method of exhaustion is sufficient for calculating $\int_a^b f(x)dx$ in the case of continuous functions since for those functions one may consider either $\sup H$ or $\inf K$. This occurs also in the case of bounded functions with a countable set of discontinuities, as it is stated in the following theorem.

**Theorem 8.3** *Let $f$ be a bounded real-valued function defined on a bounded interval $[a, b]$. Assume that $f$ is continuous or that the set of points where $f$ is discontinuous is countable. Then $f$ is Riemann integrable.*

Note that the previous theorem is a mini-version of the important Vitali-Lebesgue Theorem which states that a bounded function $f$ on a bounded interval is Riemann integrable if and only if the set of points of discontinuity of $f$ has measure zero in the sense of Lebesgue (this notion is not discussed in this book).

### 8.1.2 Antiderivatives and Fundamental Theorem of Calculus

It is known since the seventeenth century that in order to compute the Riemann integral of a continuous function $f$ it suffices to compute an antiderivative $F$ of $f$, that is a function $F$ such that the derivative of $F$ is $f$. Note that antiderivatives are also called "primitives". The precise definition is the following.

**Definition 8.4** Let $f$ be a real-valued function defined on set $\mathcal{A}$ which is an interval or a finite union of intervals (possibly unbounded) in $\mathbb{R}$. We say that a real-valued function $F$ defined on $\mathcal{A}$ is an antiderivative (or a primitive) of $f$ if $F$ is differentiable on $\mathcal{A}$ and $F'(x) = f(x)$ for all $x \in \mathcal{A}$.

A generic antiderivative of $f$, or the family of all antiderivatives of $f$, is denoted by the symbol

$$\int f(x)dx.$$

Clearly, if the domain of definition $\mathcal{A}$ of $f$ is an interval then an antiderivative is identified up to an additive constant. For example, one usually writes $\int \cos x\, dx = \sin x + k$ where $k$ denotes an arbitrary real number.

We are now ready to recall the following classical theorem.

**Theorem 8.5 (Fundamental Theorem of Calculus)** *Let $f$ be real-valued continuous function defined on a bounded interval $[a, b]$. Then the following statements hold:*

*(i) For any fixed $c \in [a, b]$, the integral function $\mathcal{I}_c$ defined on $[a, b]$ by setting*

$$\mathcal{I}_c(x) = \int_c^x f(x)dx,$$

*for all $x \in [a, b]$, is an antiderivative of $f$.*

*(ii) If a function F is an antiderivative of f on [a, b] then*

$$\int_a^b f(x)dx = F(b) - F(a).$$

Thus, establishing the existence of an antiderivative for a continuous function is not an issue, since the first part of the Fundamental Theorem of Calculus claims in particular that an antiderivative exists. In concrete problems, the main issue is to find an explicit formula for representing an antiderivative, a task which is almost prohibitive in many cases.

One of main difficulties in finding a formula for representing the antiderivative of a function consists in analysing the formal structure of the formula that defines the given function, also because the symbolic baggage, even picturesque, that history has amassed in the writing of functions, hides more than it shows. Experience shows that expressions like $\arctan x$, $\log x$, $\sqrt[n]{x}$ scare much more than those obtained by using the four operations, and more than necessary. Beyond appearance, a first useful classification of functions is that in *algebraic functions* and *transcendental functions*, see Chap. 2 for definitions. In turn, also transcendental functions can be seen sometimes as the composition of other functions of the two types. In this direction, one has to consider also the relations between the operations present in the formulas and the differentiation rules.

For example, the function $y = \frac{2x}{x^2+1}$ can be viewed as composed by a quotient, a sum and a product, but can also be thought as

$$y = \frac{1}{x^2 + 1} \cdot 2x$$

In this way one can use the chain rule

$$Df(g(x)) = f'(g(x))g'(x)$$

and get the antiderivative

$$y = \log(x^2 + 1) + k.$$

Some of the problems in this chapter aims at giving more hints on this point of view.


### 8.1.3   Idea of the proof of the Fundamental Theorem of Calculus

The main idea behind the Fundamental Theorem of Calculus is very simple. It is well known that the area of a rectangle is given by the formula

$$\boxed{\text{Area} = \text{Width} \cdot \text{Height}}$$

by which the inverse formula

$$\boxed{\text{Height} = \frac{\text{Area}}{\text{Width}}}$$

follows.

Assume now for simplicity that the function $f$ in the Fundamental Theorem of Calculus is positive. In order to compute the derivative of the integral function $\mathcal{I}_c$ at the point $x$ we have to consider the limit

$$\lim_{h \to 0} \frac{\int_c^{x+h} f(t)dt - \int_c^x f(t)dt}{h} = \lim_{h \to 0} \frac{\int_x^{x+h} f(t)dt}{h}. \tag{8.5}$$

Now, the ratio in the right-hand side of (8.5) can be understood as follows

$$\frac{\int_x^{x+h} f(t)dt}{h} = \frac{\text{Area}}{\text{Width}},$$

where the area of the mixtilinear quadrilateral $ABCD$ represented in Fig. 8.1 is at the numerator, and the corresponding width is at the denominator.

If the width $h$ of the mixtilinear quadrilateral $ABCD$ is very small, then the height can be approximated by the value of the length $|AD|$ of the segment $AD$, that is $f(x)$. Thus

$$\frac{\int_x^{x+h} f(t)dt}{h} = \frac{\text{Area}}{\text{Width}} = \text{Height} \sim |AD| = f(x).$$

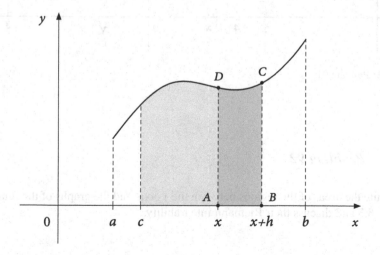

**Fig. 8.1** Geometric interpretation of the Fundamental Theorem of Calculus

In rigorous terms, this means that

$$\lim_{h \to 0} \frac{\int_x^{x+h} f(t)dt}{h} = f(x),$$

hence $\mathcal{I}_c'(x) = f(x)$ and the proof of statement (i) in the Fundamental Theorem of Calculus follows.

Once statement (i) is proved, the proof of statement (ii) easily follows by observing that any antiderivative $F$ of $f$ differs from $\mathcal{I}_c$ by a constant.

## 8.2 Problems

### 8.2.1 Problem 91

Compute the areas of the regions between the $x$-axis and the graphs of the functions in Fig. 8.2 and discuss their Riemann integrability.

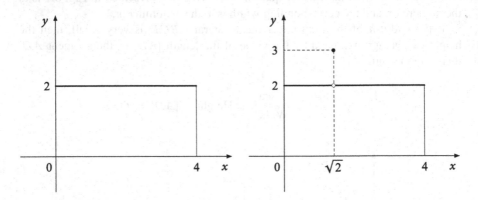

**Fig. 8.2** Problem 91

### 8.2.2 Problem 92

Compute the areas of the regions between the $x$-axis and the graphs of the functions in Fig. 8.3 and discuss their Riemann integrability.

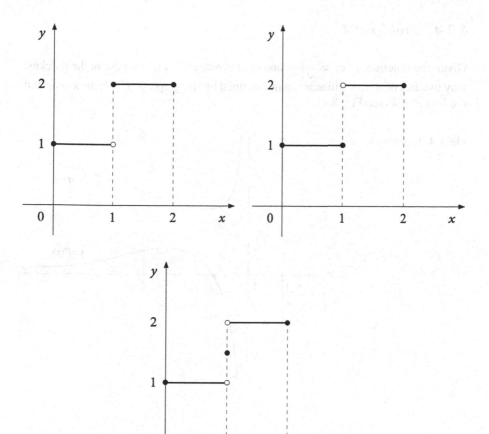

**Fig. 8.3** Problem 92

## 8.2.3 *Problem 93*

With reference to the terminology used in the answer to Problem 91, give an example of a function defined on the interval [0, 1] such that

$$\lim_{n \to +\infty} S_n > \lim_{n \to +\infty} s_n$$

### 8.2.4  Problem 94

Given the function $f(x) = \frac{x}{\sqrt{x-1}}$ and its derivative $f'(x)$, compute in the quickest way the area of the mixtilinear triangle defined by the graph of $f'(x)$, the $x$-axis and the line $x = 4$ (see Fig. 8.4).

**Fig. 8.4**  Problem 94

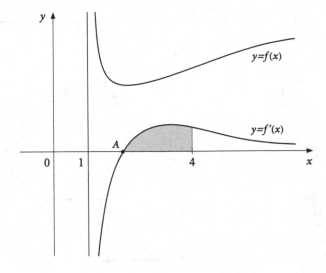

### 8.2.5  Problem 95

Compute the area of the region between the $x$-axis and the graph of the following continuous function:

$$f(x) = \begin{cases} x \log x, & \text{if } 0 < x \leq 1 \\ 0, & \text{if } x = 0 \end{cases}$$

See Fig. 8.5.

**Fig. 8.5**  Problem 95

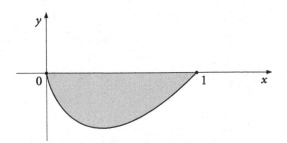

### 8.2.6 Problem 96

Given a function $f : [a, b] \to \mathbb{R}$ with the following properties:

- $f$ has an antiderivative $F$
- $f$ is Riemann integrable

prove that $\int_a^b f(x)dx = F(b) - F(a)$.

### 8.2.7 Problem 97

Given the discontinuous function

$$f(x) = \begin{cases} 2x \sin \frac{1}{x} - \cos \frac{1}{x}, & \text{if } x \neq 0 \\ 0, & \text{if } x = 0, \end{cases}$$

an antiderivative of $f$ is

$$F(x) = \begin{cases} x^2 \sin \frac{1}{x}, & \text{if } x \neq 0 \\ 0, & \text{if } x = 0. \end{cases}$$

Can we conclude on the base of the Fundamental Theorem of Calculus that

$$\int_0^{\frac{1}{\pi}} f(x)dx = F\left(\frac{1}{\pi}\right) - F(0) = 0 \ ?$$

### 8.2.8 Problem 98

Is the following proposition true or false?
The derivative of an algebraic function of degree $n$ is always an algebraic function of degree $n - 1$.

### 8.2.9 Problem 99

Say if the following propositions are true or false:

(1) The derivative of an algebraic function is always an algebraic function.
(2) The derivative of a transcendental function is always a transcendental function.
(3) The antiderivative of an algebraic function is always an algebraic function.

(4) The antiderivative of a transcendental function is always a transcendental function.

## 8.2.10  Problem 100

If a transcendental function $f(x)$ has an algebraic derivative, is there a procedure which reduces the integration of $f(x)$ to the integration of a particular algebraic function?

## 8.3  Solutions

### 8.3.1  Solution 91

In both cases, Area $= 8$ and the functions are Riemann integrable. Indeed:

**CASE A**
In this case the area of the region is just the area of a rectangle of base 4 and height 2. Moreover, the function is continuous (it is constant!) hence Riemann integrable.

**CASE B**
We divide the interval $[0, 4]$ in $n$ intervals of length $\frac{4}{n}$ by considering the finite subset of $[0, 4]$ defined by

$$P_n = \left\{ x_0 = 0, \ x_1 = \frac{4}{n}, \ldots, x_k = x_{k-1} + \frac{4}{n}, \ldots, x_n = 4 \right\}.$$

We set $s_n = s(P_n)$. Recall that $s_n$ is the area of the inscribed pluri-rectangle associated with $P_n$ and is given by the sum of the areas of the rectangles with base equal to the length of the little interval and height given by the minimum of the function on that interval. Since the minimum is always 2 in every interval, we get

$$s_n = \underbrace{\frac{4}{n} \cdot 2 + \frac{4}{n} \cdot 2 + \ldots \frac{4}{n} \cdot 2}_{n \text{ times}} = n \cdot \frac{4}{n} \cdot 2 = 8$$

Similarly, the area $S_n = S(P_n)$ of the circumscribed pluri-rectangle associated with $P_n$ is given by the sum of the areas of the rectangles with base equal to the length of the little interval and height given by the maximum of the function on that interval. In this case, the maximum is equal to 3 in one of these intervals, and is equal to 2 in the remaining $n - 1$ intervals. Thus

$$S_n = \frac{4}{n} \cdot 3 + \underbrace{\frac{4}{n} \cdot 2 + \frac{4}{n} \cdot 2 + \ldots \frac{4}{n} \cdot 2}_{n - 1 \text{ times}} = \frac{4}{n} \cdot 3 + (n - 1) \cdot \frac{4}{n} \cdot 2 = \frac{8n + 4}{n}$$

Thus the required area is computed by

$$\lim_{n \to \infty} s_n = \lim_{n \to \infty} 8 = 8$$

or, equivalently, by

$$\lim_{n \to \infty} S_n = \lim_{n \to \infty} \frac{8n + 4}{n} = 8.$$

In particular, since the two limits $\lim_{n \to \infty} s_n$, $\lim_{n \to \infty} S_n$ coincide, the function $f$ under consideration, defined by $f(x) = 2$ for all $x \in [0, 4]$ with $x \neq \sqrt{2}$ and $f(\sqrt{2}) = 3$, is Riemann integrable and

$$\int_0^1 f(x)dx = \lim_{n \to \infty} s_n = \lim_{n \to \infty} S_n = 8.$$

We note that one can deduce the Riemann integrability from the general result concerning functions with a finite set of discontinuities.

### 8.3.2 Solution 92

Area $= 3$ and functions Riemann integrable in all cases (one proceeds in the same way as in Solution 91).

### 8.3.3 Solution 93

*Example* Consider the function $f$ from $[0, 1]$ to $\mathbb{R}$ defined by

$$f(x) = \begin{cases} 1, & \text{if } x \text{ is rational} \\ 0, & \text{if } x \text{ is irrational} \end{cases}$$

See Fig. 8.6.

**Fig. 8.6** Solution 93

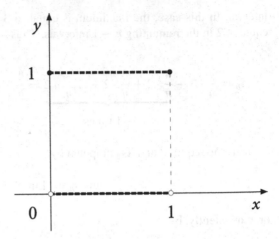

Indeed, let us divide the interval $[0, 1]$ in $n$ intervals of length $\frac{1}{n}$. We have

$$S_n = \underbrace{\frac{1}{n} \cdot 1 + \frac{1}{n} \cdot 1 + \ldots \frac{1}{n} \cdot 1}_{n \text{ times}} = n \cdot \frac{1}{n} = 1$$

and

$$s_n = \underbrace{\frac{1}{n} \cdot 0 + \frac{1}{n} \cdot 0 + \ldots \frac{1}{n} \cdot 0}_{n \text{ times}} = 0.$$

Thus

$$1 = \lim_{n \to \infty} S_n > \lim_{n \to \infty} s_n = 0.$$

We note that function $f$ is not Riemann integrable. Indeed, for any partition of $[0, 1]$ given by a finite subset $P$ of $[0, 1]$, one can prove that $s(P) \leq 0$ and $S(P) \geq 1$. This, combined with the discussion above shows that

$$\sup H = 0 < \inf K = 1,$$

where $H$, $K$ are the sets of all areas of all inscribed and circumscribed pluri-rectangles, see (8.3).

### 8.3.4 Solution 94

$$S_{ABC} = \int_2^4 f'(x)dx = [f(x)]_2^4 = \left[\frac{x}{\sqrt{x-1}}\right]_2^4 = \frac{4}{\sqrt{3}} - 2$$

*Remark* This problem was part of Question 1 of the written test in Mathematics at the so-called 'Maturità Scientifica', i.e. the graduation exam at the scientific Italian high schools in 1989. In many cases, students tried to compute

$$\int f'(x)dx = \int \frac{x-2}{2\sqrt{(x-1)^3}}dx$$

without realising that the antiderivative of $f'(x)$ is just $y = f(x)+k$. Many of them got stuck in the computations, not completely straightforward.

### 8.3.5 Solution 95

An antiderivative of $f$ is

$$F(x) = \begin{cases} \frac{x^2}{2}(\log x - \frac{1}{2}), & \text{if } 0 < x \leq 1 \\ 0, & \text{if } x = 0 \end{cases}$$

hence by the Fundamental Theorem of Calculus:

$$\int_0^1 f(x)dx = F(1) - F(0) = -\frac{1}{4}$$

and the area is $+\frac{1}{4}$.

### 8.3.6 Solution 96

*Proof* Let us consider an arbitrary partition $P$ of $[a, b]$ as in Fig. 8.7.

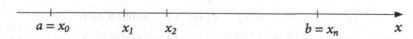

**Fig. 8.7** Solution 96

The area of the corresponding circumscribed pluri-rectangle is

$$S(P) = \sum_{i=0}^{n-1}(x_{i+1} - x_i) \sup_{x \in [x_i, x_{i+1}]} \{f(x)\} = \sum_{i=0}^{n-1}(x_{i+1} - x_i) \sup_{x \in [x_i, x_{i+1}]} \{F'(x)\}$$

where in the last equality we have simply used the fact that $F'(x) = f(x)$.

By the Lagrange's Theorem applied to the function $F$, we have that for all $i = 0, \ldots, n-1$, there exists $\xi_i \in [x_i, x_{i+1}]$ such that

$$(x_{i+1} - x_i)F'(\xi_i) = F(x_{i+1}) - F(x_i).$$

Thus, since

$$\sup_{x \in [x_i, x_{i+1}]} F'(x) \geq F'(\xi_i)$$

we have

$$S(P) \geq \sum_{i=0}^{n-1}(x_{i+1} - x_i)F'(\xi_i) = \sum_{i=0}^{n-1}(F(x_{i+1}) - F(x_i))$$

$$= (F(x_1) - F(x_0)) + (F(x_2) - F(x_1)) + (F(x_3) - F(x_2)) + \cdots$$

$$+(F(x_n) - F(x_{n-1})) = F(b) - F(a)$$

In the same way, for the area $s$ of the inscribed pluri-rectangle we get

$$s(P) \leq F(b) - F(a)$$

*Conclusion*

Let $H$ be the set of all areas of all inscribed pluri-rectangles and $K$ be the set of all areas of all circumscribed pluri-rectangles as in (8.3). For all $s \in H$ and $S \in K$ we have that

$$s \leq F(b) - F(a) \leq S. \tag{8.6}$$

On the other hand, since $f$ is integrable, we have that $\sup H = \inf K$. Summing up:

$$\begin{cases} \sup H = \inf K = F(b) - F(a), & \text{by condition (8.6)} \\ \sup H = \inf K = \int_a^b f(x)dx, & \text{by definition of integral} \end{cases}$$

Thus by the transitive property, we have

$$\int_a^b f(x)dx = F(b) - F(a)$$

### 8.3.7 Solution 97

Since $f$ is discontinuous, the Fundamental Theorem of Calculus cannot be applied directly. However, by using the improper integral and observing that the function is continuous on the intervals $[x, \frac{1}{\pi}]$ for all $0 < x < \frac{1}{\pi}$, hence the function is integrable on those intervals, we get

$$\int_0^{\frac{1}{\pi}} f(x)dx = \lim_{x \to 0^+} \int_x^{\frac{1}{\pi}} f(t)dt = \lim_{x \to 0^+} \left( F\left(\frac{1}{\pi}\right) - F(x) \right) = F\left(\frac{1}{\pi}\right) - F(0) = 0$$

Another method to prove that $\int_0^{\frac{1}{\pi}} f(x)dx = 0$ consists in using the result in Problem 96. Indeed, the function $f$ enjoys the following properties:

– $f$ has an antiderivative $F$
– $f$ is integrable (because it has only one point of discontinuity).

### 8.3.8 Solution 98

False. In fact, if we restrict our attention to polynomial functions then the property is true. For example

$$y = x^3 + 2x - 1 \quad \to \quad y' = 3x^2 + 2$$
$$y = P_3(x) \quad \to \quad y' = P_2(x)$$

In general, if $n > 0$ then

$$y = P_n(x) \quad \to \quad y' = P_{n-1}(x)$$

However, if we consider other algebraic functions, for example rational functions, we may have

$$y = \frac{1}{x^2+1} \qquad \to y' = \frac{-2x}{(x^2+1)^2}$$
$$\uparrow \qquad\qquad\qquad \uparrow$$

curve of degree 3     curve of degree 5

### 8.3.9   Solution 99

(1) True. If by algebraic function we mean any function obtained by the two elementary functions

$$y = k$$
$$y = x$$

and a finite[1] number of operations like sum, product, division, raising to a fractional power, then the proof is straightforward. Indeed:

- since the derivatives of the elementary functions

$$y' = 0$$
$$y' = 1$$

  are algebraic functions,
- since the derivatives of functions obtained by the above mentioned operations, are computed by means of a finite number of those operations applied to the functions and their derivatives

$$D(f(x) + g(x)) = f'(x) + g'(x)$$
$$D(f(x)g(x)) = f'(x)g(x) + f(x)g'(x)$$
$$D\left(\frac{f(x)}{g(x)}\right) = \frac{f'(x)g(x) - f(x)g'(x)}{g^2(x)}$$
$$D\sqrt[n]{f(x)} = \frac{f'(x)}{n\sqrt[n]{(f(x))^{n-1}}},$$

  we deduce that applying a finite number of those operations generates algebraic functions.

The proof is more difficult if we stick to the general definition of algebraic function given in Chap. 2. Indeed, if by algebraic function we denote a branch of an algebraic curve defined by an equation of the type $P(x, y) = 0$, a deeper analysis has to be carried out. Here we give some hints. Assume that $P(x, y)$ is a polynomial of the type

$$P(x, y) = \sum_{k=0}^{n} a_k(x) y^k,$$

---

[1] By using an infinite number of these operations one can obtain a transcendental function; for example: $e^x = 1 + x + \frac{1}{2}x^2 + \frac{1}{3!}x^3 + \frac{1}{4!}x^4 + \cdots + \frac{1}{n!}x^n + \ldots$

where $a_0(x), \ldots, a_n(x)$ are polynomials in $x$ with $a_n(x) \neq 0$. Assume that the given algebraic function $y = f(x)$ satisfies the equation

$$P(x, f(x)) = 0 \tag{8.7}$$

for all $x$ in some interval. Assume that the derivative $P_y$ of $P$ with respect to $y$ computed at $(x, f(x))$ is not identically zero, that is $P_y(x, f(x)) \neq 0$. Then, by differentiating both sides of Eq. (8.7) and applying the Chain Rule, we get

$$P_x(x, f(x)) + P_y(x, f(x)) f'(x) = 0$$

which implies that

$$f'(x) = -\frac{P_x(x, f(x))}{P_y(x, f(x))}.$$

Hence the function $f'$ is obtained by a finite number of algebraic operations (addition, multiplication and division) involving algebraic functions, hence it is an algebraic function. If

$$P'_y(x, f(x)) = 0 \tag{8.8}$$

then we can use Eq. (8.8) instead of Eq. (8.7) and repeat the same argument, provided the second derivative $P''_{yy}(x, f(x))$ is not identically zero. If $P''_{yy}(x, f(x)) = 0$, we can consider the third order derivative and so on. This iterative procedure will stop at least after $n$ derivatives in $y$ since the coefficient $a_n$ is not identically zero, completing the proof.

(2) False. Indeed, it suffices to consider the transcendental function

$$y = \log x$$

to see that its derivative $y = \frac{1}{x}$ (with $x > 0$) is not transcendental, i.e., is algebraic.

(3) False. Indeed, it suffices to consider the algebraic function

$$\frac{1}{x} \quad \text{with } x \in \mathbb{R} \setminus \{0\}$$

to see that its antiderivative $y = \log |x| + c$ is transcendental.

(4) True. This proposition is equivalent to (1). If $\varphi(x)$ would be an algebraic antiderivative of a transcendental function $f(x)$, then by definition of antiderivative we would have that $\varphi'(x) = f(x)$ and this would be against (1) which was proved to be true. Thus the antiderivative of a transcendental function is transcendental.

## 8.3.10 Solution 100

Yes, integrating by parts. Indeed:

$$\int f(x)dx = \int 1 \cdot f(x)dx = xf(x) - \int xf'(x)dx$$

hence the computation of the integral $\int f(x)dx$ is reduced to the computation of $\int xf'(x)dx$. The 'replacement' algebraic function is $y = xf'(x)$. Thus, using this technique allows integrating the functions $\log x$, $\arctan x$, $\arcsin x$, $\arccos x$ and functions $\log g(x)$, $\arctan g(x)$, $\arcsin g(x)$, $\arccos g(x)$ obtained by composing them with an algebraic function $g(x)$.

*Example* The replacement function of $y = \log(x^2 - 1)$ is

$$y = x \cdot \frac{2x}{x^2 - 1} = \frac{2x^2}{x^2 - 1}$$

which is not difficult to integrate.

Printed in the United States
by Baker & Taylor Publisher Services

Printed in the United States
by Baker & Taylor Publisher Services